Statistics and Computing

Series Editors:
J. Chambers
D. Hand
W. Härdle

Statistics and Computing

Antony Unwin
Martin Theus
Heike Hofmann

Graphics of
Large Datasets

Visualizing a Million

 Springer

Antony Unwin
Department of Computer
    Oriented Statistics and
    Data Analysis
University of Augsburg
Augsburg 86135
Germany

Martin Theus
Department of Computer
    Oriented Statistics and
    Data Analysis
University of Augsburg
Augsburg 86135
Germany

Heike Hofmann
Department of Statistics
Iowa State University
Ames, IA 50011-1210

Series Editors:
J. Chambers
Bell Labs, Lucent
    Technologies
600 Mountain Ave.
Murray Hill, NJ 07974
USA

D. Hand
Department of
    Mathematics
South Kensington Campus
Imperial College, London
London
SW7 2AZ
United Kingdom

W. Härdle
Institut für Statistik und
    Ökonometrie
Humboldt-Universität
    zu Berlin
Spandauer Str. 1
D-10178 Berlin, Germany

Library of Congress Control Number: 2006922760

ISBN-10: 0-387-32906-4   Printed on acid-free paper.
ISBN-13: 978-0387-32906-2

Printed in Singapore   (KYO)

9 8 7 6 5 4 3 2 1

springer.com

Preface

Analysing data is fun. It is fascinating to find out about different topics, and each dataset brings new challenges. Whether you are looking at times of goals scored in Bundesliga soccer games, ultrasound measurements of babies in the womb, or company bankruptcy data, there is always something new to be learnt. Graphic displays play a major role in data analysis, and all the main authors of this book have spent a lot of research time studying graphics and ways to improve and enhance them. We have also spent a lot of time actually analysing data. The two go together.

One of the major problems over the years has been representing data from large datasets. Initially, computers could barely read in a large dataset, so whether you could display it sensibly or not was not relevant. Gradually, computers have been able to cope with larger and larger datasets, and some weaknesses of standard data graphics have become apparent. In tackling these problems, we have become aware that there is knowledge around as to how to display large datasets, but it is not readily available, certainly not in one place. We hope that our book will help others interested in visualizing large datasets to find out more easily what has been done and to contribute themselves. More especially, we hope it will help data analysts in analysing their data.

This book grew out of discussions at a visualization workshop held in Augsburg in 2002. The main authors of each guest chapter (Dianne Cook, Ed Wegman, Graham Wills, Simon Urbanek, Steve Marron) were all there, and we are grateful to them both for agreeing to contribute to our book and for many insightful discussions at the workshop and on other occasions.

Many people have contributed to our understanding and knowledge of graphics and data analysis. Discussions at conferences, debates via email, and, especially, debates about how to analyse particular datasets have all left lasting impressions. Experimenting with software, our own and that of many others, has also taught us a great deal. We would like to thank in no particular order Günther Sawitzki, Fred Eicker, Paul Velleman,

Luke Tierney, Lee Wilkinson, Peter Dirschedl, Rick Becker, Allan Wilks, Andreas Buja, Debbie Swayne, John Sall, Bob Rodriguez, Rick Wicklin, Sandy Weisberg, Bill Eddy, Wolfgang Härdle, Junji Nakano, Moon Yul Huh, JJ Lee, Adi Wilhelm, Daniella DiBenedetto, Paul Campbell, Carlo Lauro, Roberta Siciliano, Al Inselberg, Peter Huber, Daryl Pregibon, Steve Eick, Audris Mockus, Michael Friendly, Sigbert Klinke, Matthias Nagel, Rüdiger Ostermann, Axel Benner, Friedrich Pukelsheim, Christian Röttger, Thomas Klein, Annerose Zeis, Marion Jackson, Enda Treacy, David Harmon, Robert Spence, Berwin Turlach, Bill Venables, Brian Ripley, — and too many people associated with R to name individually. Thanks are also due to former colleagues in the Statistics Department at Trinity College Dublin for provocative exchanges both over coffee and long into the night at the Irish Statistics conferences (graphical discussions, though not necessarily about graphics).

Some people read parts (or all!) of the book and made pertinent, helpful comments, but we also benefitted from encouraging and constructive criticism from Springer's anonymous referees. For help with the proofreading we are indebted to Lindy Wolsleben, Estelle Feldman, Veronika Schuster, and Sandra Schulz. Being students, Veronika and Sandra were properly respectful and careful. Lindy and Estelle called a spade a spade, especially when we had called it a spode. Our thanks to all of them; authors need both kinds of help. Needless to say (but we will say it anyway), any remaining errors are our fault. John Kimmel has been a supportive editor and it is always a pleasure to chat with him at the Springer stand at meetings. (We hope that the sales figures for our book will be good enough that he will still be prepared to talk to us in future.) We would also like to thank the six anonymous reviewers who gave significant input at various stages of the project. It takes a good editor like John to find these people.

Graphics research depends on good, if not exceptional, software to turn visualization ideas into practice. In 1987, one of us (AU) got a small grant from Apple Computers to write graphics software for teaching. (It is hard to believe, but I actually wrote some code myself initially. When I saw how good the students employed on the project were, I vowed never to write any code again.) In Dublin, it was the students Michael Lloyd, Graham Wills, and Eoin McDonnell who led the way. In Augsburg, where the Impressionists' software packages have been developed, we have benefitted from the skills of George Hawkins, Stefan Lauer, Heike Hofmann, Bernd Siegl, Christian Ordelt, Sylvia Winkler, Simon Urbanek, Klaus Bernt, Claus-Michael Müller, René Keller, Markus Helbig, Tobias Wichtrey, Alex Gouberman, Sergei Potapov, and Alex Gribov.

In 1999, John Chambers generously donated his ACM award money to the creation of a special prize for the development of statistical software by students. Recognition of the important contribution software makes to progress in research was long overdue, and we are delighted that three of our students have already won the prize.

Data analysis is a practical science and much of our knowledge and ideas stems from project collaborations with colleagues in universities and in firms. It would be impossible to talk about the problems arising from large datasets, without actually working on problems like these. We are grateful to our project partners for their cooperation and for actually sharing their data with us. It is all too easy to forget just how much work and effort is involved in collecting the data in a dataset.

Finally, as academics, we are grateful to our universities for supporting and encouraging our research, to the Deutsche Forschungsgemeinschaft for some project support, and, especially, to the Volkswagen Stiftung, whose initial funding led to the founding of the Department of Computer Oriented Statistics and Data Analysis in Ausgburg, which brought us all together.

Contents

1

Introduction

Antony Unwin

Permit me to add a word upon the meaning of a million, being a number so enormous as to be difficult to conceive.

Francis Galton, *Hereditary Genius* p.11

1.1 Introduction

Large datasets are here to stay. The automatic recording of information, whether supermarket purchase data, internet traffic, weather data or other scientific observations, has meant that there is no limit to the size of datasets to be analysed. You might ask, if a dataset is so large, why not just take a big sample? But samples will not pick out outliers, local structures, or systematic errors in the data. They will not be large enough for multivariate categorical analyses. And you have to be careful how you sample or what subset you use. In Yates's 1966 Fisher Memorial lecture (Yates; 1966), he remarked that

> *Instead, an analysis, often involving elaborate theory and requiring much numerical computation, was made on a particular batch of data, and conclusions were published. Only when similar analyses were performed on other batches of data was it found that serious contradictions existed between the different batches and that the original conclusions had to be considerably modified.*

Being Yates, he did not shrink from citing an example of a study by Fisher where this happened.

Of course, it is not always necessary or sensible to analyse all the data available. A sample of telephone calls would give a good estimate of expected revenue, whereas records of all telephone calls would be needed to spot unusual, but rare, patterns (and possible fraud). However, as datasets grow, so do subsets of datasets. There are increasingly many applications where methods are needed for coping with larger volumes of data.

The aim of our book is to look at ways of visualizing large datasets, whether large in numbers of cases or large in numbers of variables or large in both. Visualization is valuable for exploring data and for presenting information in data, whether global summaries or local patterns (to use the terminology of Bolton et al.; 2004). Data analysts, statisticians, computer scientists — indeed anyone who has to explore a large dataset of their own — should benefit from reading this book.

Why is visualizing data important? Data visualization is good for data cleaning, for exploring data — as John Tukey put it (Tukey and Tukey; 1985): "There is nothing better than a picture for making you think of questions you had forgotten to ask (even mentally)" — for identifying trends and clusters, for spotting local patterns, for evaluating modelling output and for presenting results. Visualization is essential for Exploratory Data Analysis.

Visualization is an important complement to statistical approaches and is essential for data exploration and understanding, and as Ripley (2005) says: "Finding ways to visualize datasets can be as important as ways to analyse them." There are plenty of graphical displays that work well for small datasets and that can be found in the commonly available software packages, but they do not automatically scale up. Dotplots, scatterplots, and parallel coordinate plots all suffer from overplotting with large datasets; just think of drawing a scatterplot of a million points. The number "a million" is a useful symbolic target, because visualizing a million cases already raises interesting problems and because, despite what the Galton quotation introducing the chapter says, a million cases is a comprehensible size. The UK Breast Cancer study followed a million women, there are about one million pitches thrown every five years of major league baseball, and the television quiz show, popular in many different countries, asks "Who wants to be a millionaire?" (even though exchange rate differences mean that a million is not worth the same in every country).

Graphics have played an important role in the presentation of statistical data for at least two hundred years, if Playfair (2005) is taken as the starting point. Sometimes they have been used more, sometimes less, and often statisticians have taken them very seriously. In 1973, Yates wrote in his preface to the reissue of Fisher's *Statistical Methods for Research Workers* (Fisher; 1991), referring to the first edition of 1925: "Following the introductory chapter comes a chapter on diagrams. This must have been an innovation surprising to statisticians of that time." Talking about diagrams at all still seems to surprise some statisticians.

After a period in which graphics tended to be downplayed because early computers could not produce good graphics, they have become increasingly popular in recent years. Presentation of results is one use of graphics but they are also valuable for exploring and analysing data. It is difficult to tell if this use has also increased, but it would be very surprising if it had not, thanks to the ready availability of easy to use software. Generally speaking, software for working with data is excellent for looking at small datasets. Although this is what people mostly need, it is not always enough. Firms now usually have their own data warehouse with substantial data holdings. Scientific studies can include extensive amounts of automatically recorded data. There are even many large datasets that have been made public on the web (e.g., TAO, the Tropical Atmosphere Ocean Project, or GESIS, German Social Survey Data). All of this puts heavier demands on software, and both analytic and visualization methods need to be revised and improved to deal with these new challenges.

You could argue that the challenges of dealing with large datasets are not new at all. More than a century ago, Francis Galton (1892) thought about what a million might look like. He describes in his 1869 book, *Hereditary Genius: An Inquiry into its Laws and Consequences*, counting leaves in a long avenue to give him a sense of how much a million must be (p.11):

> *Accordingly, I fixed upon a tree of average bulk and flower, and drew imaginary lines — first halving the tree, then quartering, and so on, until I arrived at a subdivision that was not too large to allow of my counting the spikes of flowers it included. I did this with three different trees, and arrived at pretty much the same result: as well as I recollect, the three estimates were as nine, ten, and eleven. Then I counted the trees in the avenue, and, multiplying all together, I found the spikes to be just about 100,000 in number. Ever since then, whenever a million is mentioned, I recall the long perspective of the avenue of Bushey Park, with its stately chestnuts clothed from top to bottom with spikes of flowers, bright in the sunshine, and I imagine a similarly continuous floral band, of ten miles in length.*

That is all very well for visualizing the idea of how much the number a million might be but does not help to visualize the characteristics of a population of a million. Being Galton, he had considered that too and in the same book (p. 28) he included a diagram showing how he imagined the distribution of the heights of a million men would look (Figure 1.1).

Galton suggested that each man would stand with his back to the wall and a mark would be made registering his height. In effect, this gives a jittered dotplot, except that there are so many dots in the centre of the distribution that there is just a solid black blob. (Incidentally, with Galton's explanation few people would have problems understanding the diagram. Would that all graphics were explained so convincingly.)

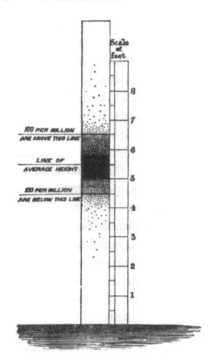

Fig. 1.1. *Galton's diagram visualizing a million in 1869.*

How would a million be visualized today? If you have ever drawn a histogram or a scatterplot of a million cases, you know that it is possible, but that there are problems. The screen resolution of a computer cannot be high enough to show very small bars in the histogram, and in regions of high density the scatterplots look like black blobs with huge numbers of points piled on top of one another. (It is noteworthy — and useful — that the weaknesses of the two kinds of plot arise at opposite extremes of the distributional densities.) So what should be visualized? If the distributional form of the bulk of the data is of interest, then the histogram will be fine for one-dimensional views (and it may give some information about outliers too). If individual outliers are of interest, then the scatterplot will be pretty good (and it will give a fair bit of distributional information as well). One aim might be described as global, attempting to summarise the main structure, and the other as local, attempting to identify individual features. Ideally, both kinds of plot are needed to satisfy both aims.

1.2 Data Visualization

Many different words can be used to describe graphic representations of data, but the overall aim is always to visualize the information in the data and so the term *Data Visualization* is the best universal term. Other terms have different connotations. At a seminar in Munich in 2004 where researchers from all fields interested in visualization met, one person thought the word 'plot' was being used to describe the story in a statistical graphic, not the graphic itself. Would that every plot had a good plot!

Deciding on which graphics to use is often a matter of taste. What one person thinks are good graphics for illustrating information may not appeal to someone else. It may also happen that different people interpret the same graphic in quite different ways. There are no absolutes like the 5% significance level for statistical tests (and whatever reservations there may be about relying on such artificial limits, they are a help in setting widely accepted standards). Buja et al. (1988) and others (e.g. Gelman; 2004) have suggested comparing a graphic by eye with sets of similar graphics generated from an appropriate model. This is too much effort to be applied in every case but can be useful in certain applications. Like most graphical methods, it relies primarily on the interocular trauma test (if it hits you between the eyes, it's significant). Results inferred from graphics should always be treated with caution. They are reassuring if they confirm something already found analytically, but they should otherwise be regarded as speculative until confirmed by other, ideally statistical, means.

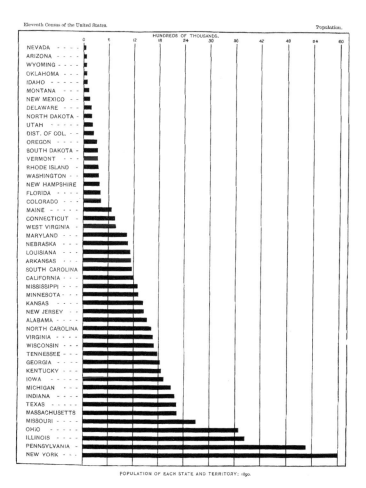

Fig. 1.2. *An early graphic of a big dataset: the US Population in 1890 from the official census report.*

Graphical displays have always been important in statistics, although they are used more by some statisticians than by others. Mostly they have been used to display data distributions, sometimes of raw data, sometimes of model residuals. A fascinating early example for a large dataset can be found in the official report on the 11th US census, which has been published on the web (http://www.census.gov/prod/www/abs/ decennial/1890.htm). It includes a horizontal barchart of the 1890 population of 63 million by State and Territory, Figure 1.2. It is fascinating to observe that California (the most populous state today) was only of middling size with a population of just over a million. As computer power has increased and computer software has become more sophisticated, more complex graphical tools have been developed and now not only data are displayed but also structures such as networks, graphical models, or trees.

It is appropriate to talk of visualization and not just of graphics. However, this does not mean the field of Scientific Visualization, where superbly realistic 3-D graphics are presented, nor does it mean Information Visualization, which is concerned with the visualization of information of all kinds (see Card et al.; 1999; Fayyad et al.; 1999; Spence; 2000, and the annual InfoVis meetings, http://www.infovis.org, for more details as well as for many intriguing and effective displays). Data Visualization is related to Information Visualization, but there are important differences. Data Visualization is for exploration, for uncovering information, as well as for presenting information. It is certainly a goal of Data Visualization to present any information in the data, but another goal is to display the raw data themselves, revealing the inherent variability and uncertainty. A provocative example of an unusual data visualization can be found in WordCount (http://www.wordcount.org) by Jonathan Harris (www.number27.org), 2004, Figure 1.3. The 86,800 different words that appear at least twice in the British National Corpus (a collection of texts comprising 100 million words in all) are displayed in order of frequency. The word 'visualization' was sought and then centered between the words one rank more frequent and one rank less frequent. The combination of the detail above and the overview below is very effective.

Fig. 1.3. *Where the word visualization comes in frequency of use amongst the texts of the British National Corpus.*

1.3 Research Literature

Research literature on Data Visualization can be found in both statistics and computer science. Statisticians are interested in data description and displays arising out of modelling data structure, reflecting statistical properties. Computer scientists' work is closely tied to information visualization. Unfortunately, as a search of the web reveals, the term Data Visualization means different things to different people. There is a book *The Art of Data Visualization* (Post et al.; 2002), which is about computer graphics (rendering issues and scientific visualization) rather than statistical graphics. The company Interactive Data Visualization produces computer games. Many software packages are available, which are capable of producing extraordinary, though not necessarily informative, displays of data.

There are several classic books on statistical graphics and each takes a different view of the topic. Tufte (1983) is full of good advice and strong opinions, pointing out the misleading effects of elaboration of statistical graphics. Cleveland (1994) concentrates on the sensible design of basic plots. Both these books have many interesting examples. Bertin (1983) attempts to provide a (French) philosophical approach, offering plenty to think about, though the practical implications are not always clear. Wilkinson (2005) describes a flexible system for implementing software for drawing statistical graphics. None of these books discusses the difficulty of visualizing large datasets, and when advice is given (e.g., in Cleveland, sunflower plots for scatterplots with overlapping points), it is of little practical help. All four authors have contributed much to our overall grasp of statistical graphics and all give good guidance for datasets that are not very large. Friendly has collected a number of excellent statistical graphics (and some awful ones to show just how bad they can get) on his web-based Gallery of Data Visualization (http://www.math.yorku.ca/SCS/Gallery/). Murrell's book (Murrell; 2005) describes how to draw all kinds of static graphics using the software R, but does not make any recommendations as to which are good or bad, nor does he discuss interpretation of graphics.

Practical statisticians have often emphasised the importance of graphical displays, but until PCs and laser printers became available in the 1980s, it was not easy to produce good graphics with a computer. It is fascinating to read how attitudes to graphics changed as statistical computing developed. While on the computational side many algorithms became practical for the first time as computers (electronic as distinct from human) took over the burden of calculation, it was difficult to produce graphics and the quality of reproduction was poor (cf. Figure 1.5).

There is a thought-provoking paragraph in Yates's Fisher Memorial lecture of 1966 on graphics:

Graph plotters are another development which is of great importance to the statistician. Computers can now readily produce diagrammatic representations of numerical material, though few of these devices are yet available in this country. They may well revolutionise the presentation of much statistical material, but their proper utilisation will require much hard thinking by statisticians.

Given the graphics displays that are to be found in much of the media nowadays, it is clear that there is still a lot of hard thinking to be done!

Chambers (1967) (the future designer of S) gave a paper at the first RSS meeting on statistical computing and posed the question "What do we need to provide in order to take full advantage of the new computers? More and better graphical techniques, surely, since these provide a more condensed summary of information on the spot." He also remarked in connection with the availability of "new input-output media (e.g., graphical displays units)" that: "There is great potential power here for data analysis but statisticians are not yet ready to exploit it." Some of us might argue that this situation has not changed a great deal in almost 40 years.

Wegman and Carr (1993) prepared a summarising review on graphics for the Handbook of Statistics. With regard to large datasets, they emphasised the limits of human perceptual ability and considered the restrictions this placed on statistical displays.

Huber wrote a review of the 1995 meeting on Massive Datasets (Huber; 1999) including the following comments on visualization. In his opinion (p. 637) visualization "runs into problems just above medium sets." This is reasonable, though pessimistic, as a medium-sized dataset would have only about 10,000 cases and 10 variables (cf. Table 1.1 later on in the chapter). Perhaps he was thinking of visualizing all individual cases, because later on in the same article (p. 646) he wrote "direct visualization of larger than medium datasets in their entirety is an unsolved (and possibly unsolvable) problem."

The most important comment in his paper concerning visualization is that "visualization for large data is an oxymoron — the art is to reduce size before one visualizes. The contradiction (and challenge) is that we may need to visualize first in order to find out how to reduce size." It is an art, because standard summarising displays like histograms or barcharts may not be appropriate or sufficient for the data in hand. In that case, micro-level or local visualization of cases may be necessary to suggest what kind of macro-level or global visualization might work.

There are not yet many papers on graphics for large datasets. One exception is Eick and Karr's (2002) on what the authors call visual scalability. They outline both the limits of human perception and the limits of the then current equipment. They demonstrate how multi-windowing, interaction, and good organisation can go a long way to displaying large datasets effectively. Another is Ward et al. (2004), which emphasises the need for multiresolution strategies, that is, representing the data at different levels of abstraction or detail. Some computer scientists have published research on visualization of large datasets and have experimented with novel ideas. Keim (2000), for instance, considers pixel-oriented ways of encoding multivariate information for large datasets.

In the United States, the National Visualization and Analytics Center (NVAC) has published a book (Thomas and Cook; 2005) on analysing large datasets for combatting terrorism. They refer (p. 4) to taking advantage of "the human eye's broad bandwidth pathway into the mind to allow users to see, explore, and understand large amounts of information at once." They seek new methods using "multiple and large-area computer displays to assist analysts" (p. 84), a different approach to the one in this book, which concentrates on improving performance on single screens, the practical situation most of us face.

1.4 How Large Is a Large Dataset?

What is meant by large, when large datasets are referred to, tends to change over time and depends on what methods are to be applied to the data. Computers have more storage and more power, and tasks that were onerous last year become run-of-the-mill this year. The Lanarkshire Milk Experiment was a large-scale study from 1930, made famous amongst statisticians by Student's devastating criticism (Student; 1931). Figure 1.4 shows the odd pattern of average growth for the 10,000 schoolgirls in the study (there was a similar pattern for the boys). Each child was measured twice, once in winter and again six months later in summer. The display shows the averages at each age and links them together to estimate an average growth curve for girls over time. Such a display would be easy to produce today, but it must have taken substantial effort seventy years ago to organise the necessary calculations.

It would be interesting to know what statisticians have thought of as large over the years, but it turns out to be difficult to pin down. There is Huber's (1992) classification, where he divided datasets from tiny up to huge based on storage space required, see Table 1.1. He estimated that a large dataset might have a million cases with 10 variables. Wegman (1995) extended the table by a further factor of 10^2 ("monster" datasets).

Once electronic computers were used more commonly in statistics, their capabilities started to determine how big the datasets that could be analysed were. There are several levels of looking at this. Firstly, you can consider the amount of data that can be stored (which is related to the Huber scale of dataset sizes and depends on the capacity of the storage media available) and identified (which depends on the software available). Secondly, you can think about what analyses can be carried out on

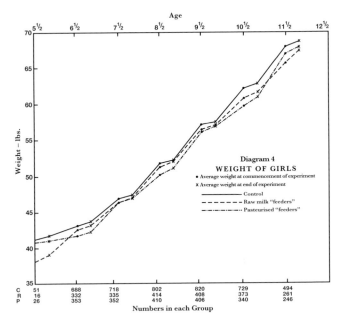

Fig. 1.4. *A reproduction of one of Student's displays from the Lanarkshire Milk Study. Notice the apparent irregularity of weight increase with age, which drew attention to some of the problems in the design of the study and in how it was carried out.*

the data. The requirements for some methods obviously grow too fast with numbers of cases (e.g., hierarchical clustering) and some software needs all the data in main memory, which automatically makes that a limiting factor. Thirdly, you can demand that analyses be carried out within an "acceptable" time. For instance, interactive analyses need very fast response times, much faster than anything dreamed about in the days when users expected to wait a few hours for their output to turn up.

A search of the major statistical journals using *JSTOR* threw up a number of comments in published papers on the size of datasets, but most

Table 1.1. *Huber's Classification of Dataset Sizes*

Size	Description	Bytes
Tiny	Can be written on a blackboard	10^2
Small	Fits on a few pages	10^4
Medium	Fills a floppy disk	10^6
Large	Fills a tape	10^8
Huge	Needs many tapes	10^{10}

were not quantified. They are listed in Table 1.2 for illustration and some are discussed in more detail in chronological order.

Table 1.2: *Quotations on Dataset Size*

Year	Comment	Author
1959	The phrase "If large-scale storage is not available" implies that a data set of 1,000 cases would have been large.	Harris
1965	"the analysis of the data recorded by Tel-Star, an early communications satellite, involved tens of thousands of observations and challenged contemporary computing technology."	Chambers
1966	"The need for better editing is well known to those concerned with extensive data sets."	Yates
1967	Datasets of "modest bulk".	Page
1967	"Vast" datasets.	Gower
1975	"It is now possible to access large data sets directly from magnetic tape."	McNeill & Tukey
1978	For SPSS "any one analytical use of the file is limited to using at most 500 variables", though up to 5,000 could be loaded in all.	Muller
1981	"There is now a collection of computer subroutines designed to summarize large data sets in histogram form." In the example he used, statistics were calculated for 20,000 samples of size 50 and histograms with 800 (!) cells were prepared for each statistic.	Dickey
1981	Restricted in their analysis at one site because the software there could only handle 88,000 real numbers.	Aitken et al.
1981	"Substantial" data sets in the census	Kruskal
1982	Moderate data sets have less than 500 cases and large have more than 2,000 (for linear-logistic models).	Koch
1986	"... allows even very large data sets to be explored interactively" and referred to a regression data set with 11,000 cases.	Gilks

continued on next page

Table 1.2: *continued*

Year	Comment	Author
1986	"The increased use of computing has in turn increased the importance of developing methods for interpretation of large volumes of data..."	Eddy
1987	"What is large depends on the frame of reference. If available plotting space for a scatterplot is a one-inch square, 500 points can seem large. For our purposes, N is large if plotting or computation times are long, or if plots can have an extensive amount of overplotting." At that time, 50,000 points was large (based on rendering time) but the authors pointed out: "The representation of 1 million or more data points in each plot is feasible."	Carr et al.
1987	"a moderate amount of data, say several hundred observations."	Becker et al.
1990	"A regression model for 5,000 cases with 6 variables would be a high sample size for immediate evaluation (c. 3 seconds), but far too big for even rough bootstrapping (estimated to take an hour)."	Sawitzki
1990	"Computing plain medians was not feasible because there were nRC = 2,621,400,000 data values in all, which could not be stored in central memory."	Rousseeuw
1991	"2% of the total census records is a very large data file". For a UK population of 65 million this would be about a million cases.	Marsh et al.
1993	"The data are listed in Good and Gaskins (1980) as a histogram of 172 bins of length 10 MeV constructed from the locations of 25,752 events on a mass-spectrum. For such a huge data set in one dimension..."	Gu
1996	"huge samples (size 100,000)" and "a large number of groups (100 say)"	Sasieni & Royston

continued on next page

Table 1.2: *continued*

Year	Comment	Author
1996	"large surveys such as the NCVS may have 60,000 or more observations, and only recently has research begun on how to plot massive data sets."	Fesco et al.
1998	A "large" data set has 3667 cases. (Scatterplots for surveys)	Korn & Graubard
1999	"We focus our attention on a very small [sic] part of the information available in these data; namely the birth weight of the 4,017,264 registered singleton births."	Clemons & Pagano

In 1959, Harris described data plotting with an IBM 650. "With a fairly small table a 650 might handle up to 1,000 non-negative observations of not over 5 digits each." The accompanying phrase "If large-scale storage is not available" implies that such a dataset of 1,000 cases would have been large. For its time, the display in Figure 1.5 must have been impressive. Of course, there is "large" and "large" and according to

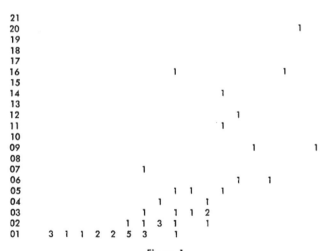

Figure 1.
21x21 scatter diagram of growth characteristics of 46 growing sub areas in the Philadelphia Region. An exponential or double-logarithmic plot reduces the scatter of this array.

Fig. 1.5. *Reproduction of Figure 1 of Harris's (1959) paper. Reprinted with permission from The American Statistician. Copyright 1959 by the American Statistical Association. All rights reserved.*

Chambers (1999), "Large-scale applications did exist, even at this time [in 1965]; the analysis of the data recorded by TelStar, an early communications satellite, involved tens of thousands of observations and challenged contemporary computing technology."

In his 1966 Fisher memorial lecture Yates wrote:

> *As an example of the type of work that can now be readily undertaken I may instance the analysis of the directional recording of swell from distant storms involving complicated spectral analyses of 10^6 automatic recordings from three pressure gauges.*

He also noted that "A serious fault of many statistical investigations in the past has been that all available data bearing on the question at issue were not made use of" — a clear cry for more powerful computing facilities to enable the analysis of larger datasets.

In 1980, Good and Gaskin published an article on what they called 'bump-hunting' and included a plot of the distribution of a dataset with just over 25,000 cases (Figure 1.6). For the 1980s, Carr et al. (1987) may sound surprisingly ambitious: "The representation of 1 million or more data points in each plot is feasible." In fact, the only surprise is that so few people have followed up on this work. Figure 1.7 shows only a quarter of a million points, but it is clear that a plot could readily have been drawn for a million. In 1995, the US National Research Council organised a workshop on "Massive Data Sets". Several of the papers were revised and reviewed a few years later and published in an issue of *JCGS* (Vol. 8, no. 3). There was some reference to numbers, but, according to Kettenring, the main organiser, in a later paper presented at the Interface meeting in 2001: "It seemed appropriate to stick with a murky definition

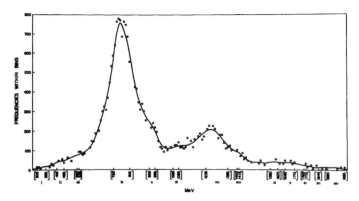

Fig. 1.6. *Reproduction of Figure A of Good and Gaskins's (1980) paper. Reprinted with permission from The Journal of the American Statistical Association. Copyright 1980 by the American Statistical Association. All rights reserved.*

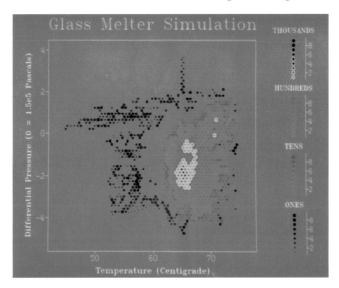

Fig. 1.7. *Reproduction of Figure 10 of Carr et al.'s (1987) paper plotting 243,800 points from a glass-melter simulation. Reprinted with permission from The Journal of the American Statistical Association. Copyright 1987 by the American Statistical Association. All rights reserved.*

[of massive]." (Kettenring; 2001) He offered one version of a murky definition as: "A massive data set is one for which the size, heterogeneity, and general complexity cause serious pain for the analyst(s)." A realistic, if unattractive, description.

In their paper, Eddy et al. (1999) worked with brain image data, analysing 2800 slices of 128 × 128 voxels each, making up about 256MB of raw data. Kahn and Braverman (1999) described climate data being collected at the rate of 80 gigabytes per day (though they did not claim to analyse datasets of this size). In a later *JCGS* paper, Braverman (2002) discussed the analysis of a subset of 5 million cases for 2 variables, i.e., large according to Huber's table. McIntosh (1999) studied telephone networks and was able to store 2 to 4 million messages on 128MB. Four years later, when he revised his paper for publication in *JCGS*, his storage limit had jumped to 55 gigabytes!

It is gratifying to be able to show a plot for a genuinely large dataset (at least in comparison to most of the datasets used so far). Figure 1.8 displays the distribution of reported birthweights for more than 4 million children. The curious form is due to rounding. (Perhaps there are more urgent matters to attend to just after a birth than to record the precise weight of the baby?) Hand et al. (2000) in their paper on Data Mining give examples of datasets that are potentially much larger than anything discussed here (Barclaycard's 350 million credit card transactions a year,

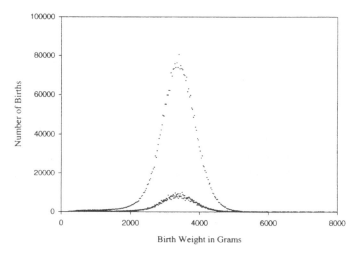

Fig. 1.8. *Reproduction of Figure 2 of Clemons and Pagano's (1999) paper. Reprinted with permission from The American Statistician. Copyright 1999 by the American Statistical Association. All rights reserved.*

Wal-Mart's 7 billion transactions a year in the early 1990s, AT&T's 70 billion long distance calls annually, reported in 1997). But have these ever been analysed in full? The examples in the paper are certainly not for small datasets (e.g., Figure 1.9), but they are not as big as that. Wilks

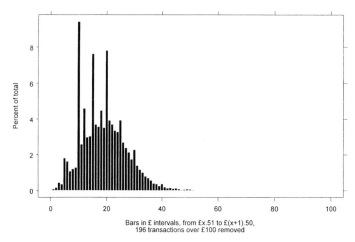

Fig. 1.9. *Reproduction of Figure 2 of Hand et al.'s (2000) paper. (Permission to reprint was granted by the Institute of Mathematical Statistics)*

talked at the 2004 Interface meeting in Baltimore about how AT&T stores details of all telephone calls on its system currently. At the moment, it is still quite difficult, but clearly it will become progressively easier. Another generator of large datasets is, of course, the internet. Marchette and Wegman (2004) report that their university network was seeing rates as high as 8 million packets per hour in 2002. As an example for large numbers of variables, consider the remark of Kettaneh et al. (2005) concerning chemometrics that "In 1970 the number of variables K (matrix columns) was large when exceeding 20. Today K is large when exceeding, say, 100,000 or 1,000,000."

For (almost) the last quotation, it is interesting to look at an article earlier than any cited above, a discussion of the analysis of the 1890 census in the US by Porter (1891). The 1890 census was the first one to count with Hollerith machines. Porter wrote "After this the 63,000,000 cards with their thousand million statements must each pass through the tabulating machine five times." Few of the other quotations come anywhere near such numbers! (We are indebted to Günther Sawitzki for suggesting looking at analyses of census data for evidence of working with large datasets.) See also Figure 1.2.

So we have had large, extensive, modest bulk, vast, substantial, huge, and massive in referring to dataset size, but rarely, probably wisely, any statements about what the terms might mean. Despite their desire for precise data to analyse, statisticians can be just as vague as the next person when it suits them. A plot of the log of dataset size against date for the quotations listed in this chapter is shown in Figure 1.10. To finish off this set of quotations, here is a final one and you might like to guess from what period it came before checking the reference:

> *Large data objects will usually be read as values from external files rather than entered during a session at the keyboard.*

Now just how large might that be? (It is actually from the 3rd edition, 1999, of the esteemed *Modern Applied Statistics with S-Plus* by Venables and Ripley. In the 4th edition, the beginning of the sentence has been changed to "For all but the smallest data sets. . . ".)

For the purposes of this book we have taken one million as a guideline, though that still leaves a lot of flexibility. It could be one million bytes (medium in Huber's classification), one million cases, one million variables, one million combinations, or one million tests. It should be something "large" at any rate.

1.5 The Effects of Largeness

Largeness comes in different forms and has many different effects. Whereas some tasks remain easy, others become obstinately difficult. Largeness is not just an increase in dataset size. Fisher's remark in his preface to the first edition of *Statistical Methods for Research Workers* in 1925 that "The elaborate mechanism built on the theory of infinitely large samples is not accurate enough for simple laboratory data" might today be re-expressed as "The elaborate mechanisms of classical statistics for analysing small samples of simple laboratory data are not enough for large, complex datasets."

Areas affected by largeness include storage, data quality, dataset complexity, speed, analyses, and, of course, visualization.

Large will usually mean large rather than LARGE in this book. It will be assumed that the whole dataset resides on the local hard disk. Problems of retrieving data from distant databases in real-time do not occur in this case. This is a restrictive assumption but may be justified on two grounds: many large datasets are small enough to be handled locally (storing a million cases on a laptop is no big deal) and methods of visualizing datasets of this size are still not fully developed.

1.5.1 Storage

The most obvious effect on increasing dataset size is the storage space needed. Although storage was a major limitation in the past (cf. McNeill and Tukey; 1975), it is far less so now. AT&T may have organisational problems in storing the information on all the calls that go through its networks (cf. Wilks's talk at the Interface meeting 2004), but AT&T can do it. For the rest of us, it is a case of being able to put on our laptops datasets of a size that only specialist installations could handle a few years ago.

Storing data is one side of the coin and retrieving data is the other. Modern database systems give intelligent access to large amounts of data,

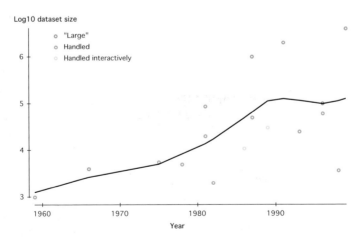

Fig. 1.10. *A scatterplot of the dataset sizes reported in the quoted papers.*

but how can the data be inspected and reviewed? Summaries and displays are all very well, but often it is valuable to look at the raw data. Standard statistics software packages offer spreadsheet-like data tables to view datasets. Whether the data can be stored locally or not, scrolling through tables is not effective. Perhaps this is the reason why Excel limits itself to a rather miserly 65,536 rows (although why only at most 256 columns are permitted remains a mystery). Tables are fine for viewing sections of a dataset, but simple scrolling is no longer a practical navigational option.

Whereas the question of storing the actual data is one problem, another difficulty can arise when converting data between different applications with different (file) formats. Most analysis or visualization tools have their own native storage format or rely on the generality of flat ASCII files. Efficient import functions are crucial, as converting data can turn out to be a major bottleneck.

1.5.2 Quality

The larger the dataset the more errors there are. Probably there is a higher proportion of errors too. As Yates (1966) put it: "The need for better editing is well known to those concerned with extensive datasets." What he meant by "extensive" in 1966 is not recorded, but as an example he gives a dataset Healy reported on in 1952 with around 4,000 cases of 30 variables each. Ripley (2005) wrote in relation to car insurance proposals that the more questions potential customers are asked the less reliable their answers will become. Huber (1994) referred to "the rawness of raw data" and wrote: "We have found that some of the hardest errors to detect by traditional methods are unsuspected gaps in the data collection (we usually discovered them serendipitously in the course of *graphical checking* [our italics])."

In addition, the larger the dataset the more coding problems there are (varieties of missings, and coding variations in general, as the dataset will often have been compiled from different sources). Data cleaning becomes a major issue — or rather it should become a major issue. Although many organisations go to considerable effort to ensure the quality of their data, many large datasets are still presented in a lamentable condition, as if the effort and expense of collecting them justifies neglecting any kind of serious preparatory cleaning. Of course, data analysts have often been given unchecked datasets to analyse, but with small datasets it is a simple job to fix the blatant errors and further errors can be checked and discussed with the dataset owner. This may sometimes lead to a gain of background information that might otherwise not have occurred. That kind of interaction with a domain expert is much more difficult to achieve when the dataset is large and the results of many individuals' efforts. No one person may have a complete overview and you cannot expect to discuss large numbers of individual cases. What all practising data analysts agree on is that the proportion of project time spent on data cleaning is huge. Estimates of 75–90% have been suggested.

Matters do improve with time in some respects. The famous curiosity of the Indians and the teenage widows from the 1950 US census (Coale and Stephan; 1962) would probably have been identified and corrected much more quickly with modern interactive tools.

1.5.3 Complexity

Largeness may mean more complexity — more variables, more detail (additional categories, special cases), and more structure (temporal or spatial components, combinations of relational data tables). Again this is not so much of a problem with small datasets, where the complexity will be by definition limited, but becomes a major problem with large datasets. They will often have special features that do not fit the standard case by variable matrix structure well-known to statisticians. Datasets of results from the PISA study comparing schoolchildren's educational achievements form an interesting example. The sample design used is weight-adjusted at many different levels to ensure scientific comparability but at the expense of transparency and understanding (Adams and Wu; 2002).

1.5.4 Speed

Largeness affects speed. It takes longer to read and retrieve data (though locally stored datasets should not have this problem), and it takes longer for some calculations to be evaluated. Formerly, the rendering of displays would also take noticeably longer, though this seems barely an issue nowadays. But it is not just the computer's speed that is affected, so is the user's. There are likely to be more variables to consider, more displays to manage, and more results to analyse. More thought is needed between the steps of an analysis, and more time is needed even just to locate objects. Locating one variable out of three or finding two cases out of seventy is easy. When there are two hundred variables and one million cases, both of these tasks require highly organized data, and software support to match, to be carried out at all.

Speed is relative. The US Census of 1890 was concerned about the analysis taking longer than 10 years (the 1880 results were first ready in 1888). Nowadays, interest lies in results fast enough to be considered interactive.

1.5.5 Analyses

Largeness implies more analyses and more results. Managing the process of analysis successfully becomes a central task in its own right. The days of typing "Statistics all" and picking up a few pages of computer output are long gone. Even in 1967, serious analysts, (Page et al.; 1967), were getting bogged down in paper: Healy p. 136, "One final thing I would press for [...] is more use of graphical output, notably destructible graphical output which does not leave heaps of paper lying around when a job is finished." The days of trawling through endless volumes of frequency tables for every variable and of contingency tables for every pair of variables are still sadly with us. Automatic filtering and storing of results are essential first steps to help analysts to concentrate on the important issues that require human input to interpret the results. This kind of work is currently more likely to be carried out under the name Data Mining than under the name Statistics.

Data Mining is mainly concerned with the analysis of large datasets (Hand et al.; 2000, mention several extremely large datasets), although some Data Mining papers only describe working with relatively small ones. For instance, the 2003 review of Data Mining software packages (Haughton et al.; 2003) used two datasets, one with 1,580 cases and one with just under 20,000 (though it did have 200 variables). Generally speaking, there is less concern in Data Mining for statistical concepts and more attention paid to computational efficiency and heuristic approaches. In this way, Data Mining and Statistics complement each other well.

1.5.6 Displays

Largeness affects screen real estate. The more cases and the more points or bars there are to be displayed, the larger a window tends to have to be (cf. Carr et al.; 1987). Visibility is obviously an issue for numbers of cases, but numbers of variables and numbers of displays are a bigger problem. The larger the dataset, the more displays are likely to be required, both for the larger number of variables and for the larger number of subsets that may need to be examined. More and bigger plots mean that window design and window management become increasingly important. In the DDB Needham dataset discussed in Putnam's (2000) *Bowling Alone* book, there are more than 20 variables relating to the number and age of children in the household. Displaying barcharts of all these variables simultaneously on a screen of 1024×768 pixels would allow a maximum of about 200×180 pixels per plot. Positioning the displays in informative ways, for instance, placing all the variables on ages of the children in a line in order, would require additional capabilities. Repositioning, reordering, and resizing graphics is difficult, messy work, so it is only done if it absolutely has to be. Like many other operations, there is a lot you might want to do with advantage, if you only had the tools to do it. These are tasks the computer can do much better than a human — provided that you can tell it what you want.

1.5.7 Graphical Formats

Dealing with large datasets is also a challenge for the purely technical process of plotting the data on the screen and saving the resulting graphics to a file. A scatterplot of Fisher's Iris Data can be plotted by whatever graphics engine you like and saved into any file format. This is no longer true when datasets become really big. Plotting hundreds of thousands of points on the screen is slow, unless the system can take advantage of the native routines of the graphics board of the computer, today usually some kind of OpenGL. This is especially true when using smoothing and sub-pixel-rendering of the plotted objects.

Similar problems arise when saving a graphic to a file. Object oriented formats like PS or PDF would be the preferred option, but saving tens of thousands of polygons of a parallel coordinate plot to a pdf-file makes this file too big for most of the tools available. At this point, bitmap graphics formats like PNG or JPEG are a more efficient choice, though they offer poorer reproduction quality. Much care must then be taken to achieve a high rendering quality on the screen.

For reasons of size, a few of the graphics reproduced in this book are "only" bitmap graphics.

1.6 What Is in This Book

This chapter has set the scene for what our book is about. The next chapter defines and discusses standard statistical graphics. Standard means the plots that are used most frequently in data visualization and that cover the basic data display requirements: barcharts, histograms, scatterplots, and so on. It is helpful to set down what is meant in this book by these terms, as not everyone agrees on the definitions of the basic plots. There is a particularly varied set of definitions in use for boxplots — sometimes more than one can be found in the same software package! Many plots can be used in data visualization: piecharts, rose diagrams, starplots, stacked barcharts, to name only a few, but the ideas described in this book can be applied to all of them just as to the standard displays.

Chapter 3 looks at the issue of upscaling graphics to cope with large datasets. In general, area plots can be upscaled with minor amendments, whereas point plots require more substantial changes.

One of the ways of extending what can be achieved with statistical graphics is to add interaction. In the fourth chapter, several developments of interactive methods are explained, which improve the capability of graphics for finding information in large datasets.

After these three chapters on general principles follow chapters on particular types of graphics for applications. These chapters have been written by experts in their field. Some have given a complete overview; a couple have chosen to concentrate on special issues.

Chapter 5 discusses specialist plots for multivariate categorical data — mosaic plots and their variations. For mosaic plots, features like sorting and redmarking are most valuable. Chapters 6 and 7 consider multivariate continuous data, firstly experimenting using pre-calcuated videos to enable selection and linking in grand tours for very large datasets and secondly looking at parallel coordinate plots for continuous data. For parallel coordinate plots, sorting is also important, but Chapter 7 concentrates on transforming link functions between the axes and using saturation brushing.

The emphasis up till now has been on visualization of cases from large datasets, but the next two chapters discuss visualization of structures — networks and trees. Like parallel coordinates, networks are drawn with many lines, and so an increase in magnitude has a more dramatic effect on networks than it does on point or area plots. The main issue is not drawing optimal layouts but drawing informative and acceptable layouts fast enough to be useful. In particular, this chapter makes clear that having to analyse applications with a million nodes is not at all unusual. With trees, the task is different again. Large datasets do not lead to specially large trees, but complex datasets may lead to many, many trees, and the visualization here concentrates on the task of combining and summarising the information from large numbers of trees. A broad range of innovative displays is introduced for these specialist tasks, though they all have their origins in existing plots.

There are several applications discussed throughout the book. Chapter 10 looks at a major one that everyone meets in their daily work even if they are only subliminally aware of it: internet packet data. The problems of sampling the data to produce representative displays are highlighted and the aptly named 'mice and elephants plot' is shown to have good properties for uncovering features in the data.

No matter how interesting the papers are that are written by researchers to investigate deep theoretical problems, readers usually appreciate some clear-cut, cogent advice. The final short chapter uses an example to illustrate how a large dataset can be explored with visualization. If you have an immediate data visualization problem for a large dataset in front of you, the last chapter might give you some ideas.

1.7 Software

At various points in the book, several different software packages are used or referred to. Some authors give short lists of tools that implement the concepts they discuss. The packages used to produce the graphics in the book (in alphabetical order) were Data Desk, GGobi, ExploreN, KLIMT, Limn, MANET, Mondrian, R, VizML. Current references to these can be found on the book's webpage. Many of the graphics could have been produced by Excel, JMP, SAS-Insight, SPSS, S-Plus, Xlisp-Stat, or indeed by many other packages.

It is the ideas and concepts that are important and not the current implementations and so no specific recommendations are made. Everyone should use the software that they feel comfortable with and they should demand some, if not all, of the features expounded in this book. Of course, experimenting with new tools broadens the horizon and encourages the generation of new ideas. As always, we should hope that much better software is just around the corner. Every now and again that turns out to be true.

1.8 What Is on the Website

Although the book is a stand-alone resource, the accompanying website will offer additional material and information. The website at `http://www.rosuda.org/GOLD` will include:

- pdf-files and code/settings for figures;
- up-to-date links to the software mentioned;
- the most important datasets used in the book;
- errata.

1.8.1 Files and Code for Figures

Visualizing large datasets almost always involves some non-trivial graphics. If possible, the code and settings used to produce the graphics will be included on the webpage so that readers can replicate them.

Colour printing helps a lot to cope with the complexity of some graphics, but a good choice of colour schemes is important. Not all the book's graphics are optimal in this respect. Readers may like to experiment with redrawing them themselves using other colour palettes.

Many displays in the book would need far more printing space than the page size allows. Therefore, all graphics of the book can be downloaded from the website as pdf-files for closer scrutiny (as long as no copyright applies).

1.8.2 Links to Software

The book communicates principles, not details of specific implementations. Nevertheless, even the best considered principle needs an implementation in software and evaluation by users. The website will give up-to-date links to the software used in the book and, for each tool, a list of the figures that were directly drawn by that software (though in many cases other software could have been used just as well).

1.8.3 Datasets

The index lists around 20 different datasets used in the book. Some of them are used only once; others are used throughout the book. Some have millions of observations; some are smaller. The website will offer links to all datasets mentioned in the book as far as possible. The following section briefly describes seven of the datasets.

Small Datasets for Illustration

The first part of the book illustrates many principles that can best be introduced with datasets of small size.

- **Italian Olive Oils**
  The data consist of 572 samples of Italian olive oils. For each sample, the contents of the 8 major fatty acids have been determined. The data are grouped according to a hierarchical classification of 3 regions and 9 areas of Italy.

- **Athens Decathlon 2004**
  For the 28 athletes from 22 countries who completed all 10 disciplines in the 2004 Olympics decathlon in Athens, the individual results of the disciplines and the resulting point scores have been recorded.

- **2004 Cars Data**
  Taken from the website of the *Journal of Statistical Education* at `http://www.amstat.org/publications/jse/jse_data_archive.html`, this dataset contains information on a number of cars including horsepower, mileage, weight, and country of origin.

Large Datasets

Large datasets often have a very specific structure. The following four examples cover a broad range of applications.

- **Bowling Alone**
  This dataset is one used in Robert Putnam's book *Bowling Alone*. The DDB Life Style Survey, available on the web from `http://www.bowlingalone.com`, is an annual survey over 24 years of around 3,500 different individuals each year with up to just under 400 pieces of information per case. With 85,000 cases this means that, ignoring

missing values, there are about 30 million pieces of information in the dataset.

- **Bank Deals**
  Over two years there were approximately 700,000 transactions carried out for firms by a major bank. The record of each deal included the amount, the book profit, the type of deal, the location, and various other pieces of information. Recordkeeping changed between the two years so that the data are not fully comparable for the two years. These data are confidential.

- **US Current Population Survey – Census Data**
  The dataset contains data for the 63,756 families that were interviewed in the March 1995 Current Population Survey (CPS). These include families with husbands and wives, whether military or non-military; families with male heads only; families with female heads only; non-family householders and unrelated individuals are included too — although such individuals do not constitute "families" in the strict sense of the word. This dataset also includes families in group quarters. The source of the data is the March 1995 Current Population Survey, conducted by the US Bureau of the Census for the Bureau of Labor Statistics. The data can be found at `http://www.stat.ucla.edu/data/`.

- **Internet Usage Data**
  These data come from a survey conducted by the Graphics and Visualization Unit at Georgia Tech October 10 to November 16, 1997. Details of the survey are available at `http://www.cc.gatech.edu/gvu/user_surveys/survey-1997-10/`. The particular subset of the survey used here is the "general demographics" of internet users. The full survey is available from the website above, along with summaries, tables, and graphs of their analyses.
  These data were used in the American Statistical Association Statistical Graphics and Computing Sections 1999 Data Exposition.

1.9 Contributing Authors

Antony Unwin is Professor of Computer Oriented Statistics and Data Analysis at the University of Augsburg.

Martin Theus is a senior researcher at the University of Augsburg.

Heike Hofmann is Professor at Iowa State University.

Dianne Cook is Professor at Iowa State University. Her work is joint research with Peter Sutherland, Manuel Suarez, Jing Zhang, Hai-Qing You, and Hu Lan and was supported by funding from National Science Foundation grant #9982341.

Leslie Miller is Professor at Iowa State University.

Rida Moustafa completed this work on a postdoctoral appointment at the Center for Computational Statistics, George Mason University, USA. His work was funded by the Air Force Office of Scientific Research under contract F49620-01-1-0274.

Ed Wegman is Director, Center for Computational Statistics, George Mason University, USA. His work was funded by the Office of Naval Research under contract DAAD19-99-1-0314 administered by the Army Research Office, by the Air Force Office of Scientific Research under contract F49620-01-1-0274 and contract DAAD19-01-1-0464, the latter also administered by the Army Research Office, and finally by the Defense Advanced Research Projects Agency through cooperative agreement 8105-48267 with the Johns Hopkins University.

Simon Urbanek is in the research department of AT&T, New Jersey.

Graham Wills is Principal Software Engineer at SPSS, Chicago.

Bárbara González-Arévalo is at the University of Louisiana.

Félix Hernández-Campos is at the University of North Carolina.

Steve Marron is at the University of North Carolina. His research was partially supported by Cornell University's College of Engineering Mary Upson Fund and by NSF Grants DMS-9971649 and DMS-0308331.

Cheolwoo Park is at the University of Florida.

Part I

Basics

2

Statistical Graphics

Martin Theus

Millions of stars that seemed to speak in fire.

John Masefield, *The Wanderer*

2.1 Introduction

Statistics has its own basic suite of domain-specific visualization tools. These statistical graphics can best be classified by the kind of data that they depict. Statistical data are usually characterized by their scale: nominal, ordinal (which are both categorical) or numerical (which is usually regarded as continuous). What is most important in distinguishing statistical graphics from other graphics is their universality: statistical graphics are not tailored towards only one specific application but are valid for any data measured on the appropriate scales.

Depending on the data scale, certain standard graphing techniques have prevailed. Data on a discrete scale, which represent counts of different groups within a dataset, are best represented by areas whose sizes are proportional to the counts they represent, whereas data on a continuous scale are usually depicted by a single glyph (i.e., a graphical object, usually a point) per observation. Exceptions are plots based on summaries of the data like histograms and boxplots.

Using this classification, the standard statistical graphics are introduced in the following sections along with their most common extensions and modifications. Obviously the range of statistical graphics is much broader than the basic plots that can be introduced in this chapter.

2.2 Plots for Categorical Data

Plots for categorical data have not received much attention in the past. This can be explained by several factors. One reason was that software was not able to read non-numeric values as class labels, and once the

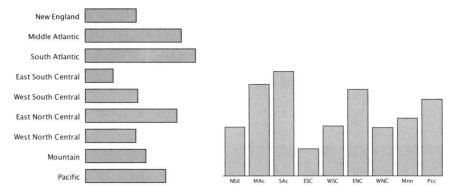

Fig. 2.1. *Two barcharts for* Division *in the* Census *dataset.*

labels had been converted to numbers, it was quite tempting to treat the data as numerical. Another reason was the difficulty of representing multivariate categorical data.

2.2.1 Barcharts and Spineplots for Univariate Categorical Data

A univariate vector of categorical data can be summarized in a one-dimensional table. The easiest way to depict this summary is a barchart, where the area of the bar represents the count for its category. For ease of comparison, all bars are given equal width, so that their heights represent the counts as well. (The bars can also be drawn horizontally with equal heights but different lengths.)

Figure 2.1 shows an example of two barcharts. In the plot on the left the bars are plotted horizontally, and on the right the bars are drawn vertically. Whereas the vertical plot follows the more natural stacking principle, the horizontal one is better for labelling the graphic.

Spineplots

In many situations it is desirable to look at the distribution of a subgroup of a categorical variable. Going back to the example of Figure 2.1, it might be interesting to know the proportions of females in the different divisions. Figure 2.2 shows two examples of how this extra information may be included in a barchart. The left barchart just adds a barchart for females within the same scale of the barchart for the complete sample. The right plot is what most spreadsheet-like applications do: separate barcharts for each group are plotted side by side for all divisions. This plot is much harder to read, as there are twice as many bars as before. However, the left plot does not make comparing the proportions of females across the different divisions easy. Each bar of a subgroup must

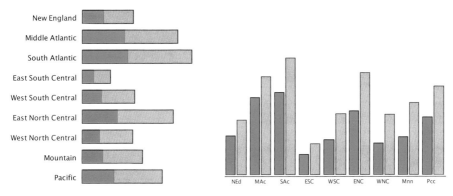

Fig. 2.2. *Two barcharts showing the proportion of* females *(red highlighting) in each* Division *in the* Census *dataset.*

be compared to the bar of the complete sample, which is visually not a straightforward task. A solution for this problem is the spineplot. In a spineplot all bars have equal length, but proportional width, which retains the proportionality of the area. As the length of each bar is now standardized to be 100%, the highlighting can be drawn from 0% to the proportion of the subgroup in a particular bar. Figure 2.3 illustrates a highlighted barchart and spineplot. Whereas the barchart enables a good comparison of the absolute counts of the subgroup, the spineplot enables a direct comparison of the proportions.

2.2.2 Mosaic Plots for Multi-dimensional Categorical Data

Highlighted spineplots already incorporate information of more than just one categorical variable in a single plot. Mosaic plots, which were first

Fig. 2.3. *A spineplot (right) allows the comparison of proportions across categories.*

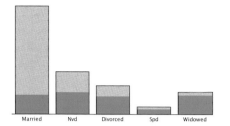 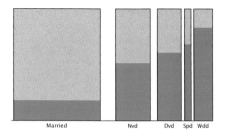

Fig. 2.4. *A barchart and spineplot for marital status with female heads of household selected.*

introduced by Hartigan and Kleiner (1981), are designed to show the dependencies and interactions between multiple categorical variables in one plot. Figure 2.4 shows a barchart and spineplot for marital status with female heads of household selected. A spineplot can be regarded as a kind of one-dimensional mosaic plot. Figure 2.5 shows the corresponding mosaic plot for the data in Figure 2.4. In contrast with a barchart, where the bars are aligned to an axis, the mosaic plot uses a rectangular region, which is subdivided into tiles according to the numbers of observations falling into the different classes. This subdivision is done recursively, or in statistical terms conditionally, as more variables are included. Figure 2.6 shows an example of a mosaic plot for the US-Census data. Starting with the variable *Marital Status*, the initial rectangle (i.e. the complete square) is divided along x according to the classes of this variable (cf. Figure 2.6, left). In the second step, the variable *Education* — summarized as a binary response of *college education or higher* and *highschool or less* — is incorporated. This is done by dividing each bar or tile of *Marital Sta-*

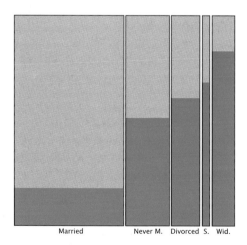

Fig. 2.5. *A one-dimensional mosaic plot for marital status*

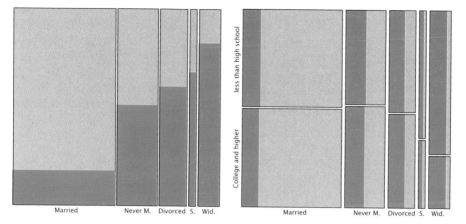

Fig. 2.6. *Development of a mosaic plot including* Marital Status *and* Education *(all females are highlighted).*

tus according to *Education* along y. The plot (Figure 2.6, right) shows that the groups of people who are married or were never married have approximately equivalent education levels. Only about one fourth of all widowed heads of household have some college education. The third variable in the plot is *Gender*, which is shown as horizontal highlighting in all combinations of *Marital Status* and *Education*. This plot shows that *Gender* and *Education* are independent, given any class of *Marital Status*. At this point it is easier to use highlighting instead of including the gender information via a third variable as the next split along x.

A detailed discussion on how to detect dependence structures in mosaic plots can be found in Theus and Lauer (1999).

The recursive construction of a mosaic plot means that the only limit for the number of variables included is the number of tiles to display, i.e. the number of possible combinations of the variables. Labeling of mosaic plots is not an easy task even with relatively few variables. Usually it is possible to label the first 2 to 3 variables in the plot all further information should be provided by interactive queries (cf. Chapter 4). If interactive queries are not available, the following strategy has proved to be helpful. Variables with only few categories should be put in the plot first, to keep the number of conditioned groups small. If one of the variables in the plot is a binary response, showing this variable via highlighting will reduce the number of tiles by half.

Note that the gaps between the tiles are not part of the rectangular region that is used to build the tiles. The gaps are there to improve visual discrimination.

2.3 Plots for Continuous Data

The most commonly used plots for continuous data are dotplots, box-plots, and histograms for one-dimensional data and scatterplots for two-dimensional data. Methods and plots for higher dimensions of continuous data include parallel coordinates and the Grand Tour.

2.3.1 Dotplots, Boxplots, and Histograms

The simplest way to plot univariate continuous data is a dotplot. Because the points are distributed along only one axis, overplotting is a serious problem, no matter how small the sample is. The usual technique to avoid overplotting is jittering, i.e., the data are randomly spread along a virtual second axis. Figure 2.7 shows an example of a jittered dotplot for the variable *Horsepower* of the Cars2004 data. Although jittering was used, the overplotting still masks parts of the distribution, and a quantification of the density is not easily possible. Nevertheless, the accumulation of cases around 300 hp and the gap between 400 hp and 450 hp are clearly visible, which might not be the case in other plots.

Histograms use area to represent counts of a distribution. This makes them somewhat related to barcharts and mosaic plots, although the number or the width of the bins of a histogram is not determined a priori and the bins are drawn without gaps between them reflecting the continuous scale of the data. Whereas barcharts and mosaic plots show the exact distribution of the sample, a histogram is always just one approximation to the distribution of the data. Sometimes histograms are also used as crude density estimators for some "true", but usually unknown, underlying distribution for the data. There are much better density estimation methods that produce smooth distribution displays. Average Shifted Histograms (ASH) (Scott; 1992) is one of them — see also Section 6.2. Figure 2.8 shows a histogram for the same data as in Figure 2.7. The histogram

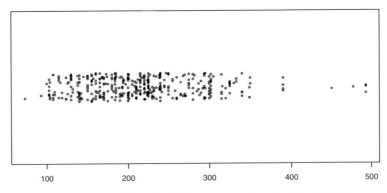

Fig. 2.7. *A jittered dotplot of* Horsepower *for the Cars2004 data.*

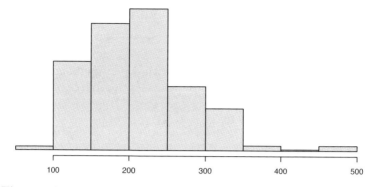

Fig. 2.8. *A default histogram of* Horsepower *for the Cars2004 data.*

gives a much better impression of the distribution than the dotplot in Figure 2.7 does. A good approximation to the shape of the data distribution depends strongly on the chosen anchorpoint, i.e., the point where the first bin starts, and the binwidth, i.e., the width of the intervals used to draw the histogram. For the example, the accumulation of cases at about 300 hp is not visible in Figure 2.8. Changing the anchorpoint from 50 to 70 and the binwidth from 50 to just 20 gives the plot in Figure 2.9, which now shows all the features found in the dotplot in Figure 2.7. Being able to change these plot parameters quickly is a typical interactive feature, which will be discussed in more detail in Chapter 4.

Boxplots are another popular alternative for univariate continuous data. Boxplots use a mixture of summary information (as histograms do) and information on individual points (as dotplots do). The center line in a boxplot is marked by the median $\tilde{x}_{0.5}$ of the sample. Upper and lower ends of the box are determined by the upper and lower hinges x_u and x_l, which

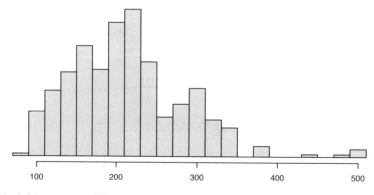

Fig. 2.9. *A histogram of* Horsepower *for the Cars2004 data, with anchorpoint 70 and binwidth 20.*

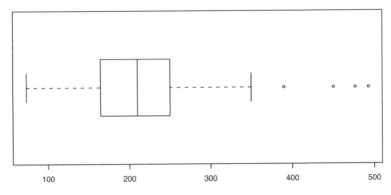

Fig. 2.10. *A boxplot of* Horsepower *for the Cars2004 data.*

are the medians of the upper and lower halves of the sample divided at the median. The dotted lines — the so called "whiskers" — are drawn from the box up to the most extreme point, which is not further away from the hinge than 1.5 times the h-spread (defined as the distance between the hinges), i.e., $x_u - x_l$. All points further away from the center are marked as outliers by a single dot. If points exceed 3 times the h-spread, they are marked as extreme-outliers. The virtual thresholds, hinge + 1.5 or 3 times the h-spread, are called inner fences and outer fences.[1]

Figure 2.10 shows the data from Figure 2.7 and 2.8 in a boxplot. The boxplot can neither show the second mode around 300 hp nor any gaps in the data that are not between outliers. Boxplots are very good for comparing distributions because they take up little space and can be drawn parallel to one another.

To highlight a subgroup of the data in a boxplot, the boxplot for all data is often modified as shown in Figure 2.11. In the base boxplot showing all data, the whiskers are drawn as light-gray boxes, which allows plotting of a standard boxplot for the highlighted data on top of the base boxplot for all data.

Table 2.1 gives an overview of the strengths and weaknesses of the three plots.

Fig. 2.11. *A boxplot of* Horsepower *for the Cars2004 data, with all 4-cylinder cars highlighted.*

---

[1] Many statistical software packages do not stick to Tukey's original definition, which can make boxplots confusing to interpret.

Table 2.1. *A Comparison of the Strengths (+) and Weakness (–) of Plots for Univariate Continuous Data ('○' Means Neither Strength nor Weakness)*

	Dotplot	Histogram	Boxplot
Visualizing the shape of a distribution	–	+	○
Detection of outlier	+	–	+
Inspection of gaps, discreteness	+	○	–
Size of the sample	○	○	–
Comparison of distributions	–	○	+

2.3.2 Scatterplots, Parallel Coordinates, and the Grand Tour

Except for the dotplot, none of the plots presented in the last section has a natural generalization to more than one dimension. The scatterplot is the natural counterpart of a dotplot in two dimensions — the 3-D rotating plot in three dimensions.

Scatterplots and Scatterplot Matrices

The scatterplot is the ideal plot to display the structure of two-dimensional continuous data. In a scatterplot, two variables are plotted in a cartesian, i.e., orthogonal coordinate system.

Higher-dimensional structures are often depicted in a so-called scatterplot matrix or SPLOM. In a SPLOM for k variables, $\binom{k}{2}$ scatterplots are plotted to display the $k(k-1)/2$ bivariate relationships of all k variables. The scatterplot of variable i vs. variable j is plotted in the upper triangle matrix of size $k-1$ at position (i,j). Figure 2.12 gives an example of a scatterplot matrix for 5 variables. The 10 scatterplots are arranged to accommodate all 10 pairs of the 5 variables. Often a univariate plot of variable i is plotted at position (i,i) in the plot matrix. In Figure 2.12 only the names of the variables are noted on the diagonal, and the 10 transposed plots are shown in the lower triangle matrix.

The natural generalization of a two-dimensional scatterplot is the three-dimensional rotating plot, which gives a three-dimensional view of continuous data. The pseudo rotation is generated by a rapid succession of two-dimensional projections with smoothly changing projection angles.

The Grand Tour

So far, all plots have been rendered on a two-dimensional medium, a sheet of paper or a computer screen. Even a 3-D rotating plot is "just" a 2-D

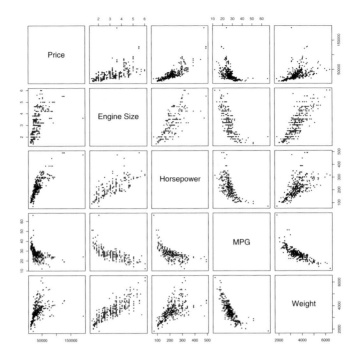

Fig. 2.12. *A full scatterplot matrix for 5 variables from the cars dataset.*

projection of 3-D data, which gets its pseudo three-dimensionality only by the "motion" of the rotation. The *Grand Tour* as introduced by Asimov (1985) generalizes the idea of a 3-D rotating plot to an arbitrary number of dimensions. It is defined as:

> A continuous 1-parameter family of d-dimensional projections of p-dimensional data which is dense in the set of all d-dimensional projections in $I\!R^p$. The parameter is usually thought of as time.

For a 3-D rotating plot, the parameter p equals 3 and the parameter d equals 2. In contrast with the 3-D rotating plot, the Grand Tour does not have rotational controls, but uses successive randomly selected projections. Figure 2.13 shows an example of 3 successive planes $P1, P2$, and $P3$ in three dimensions. The planes between the randomly selected base planes are interpolated to get a smooth pseudo-rotation, which is comparable to a 3-D rotation. A sample projection plane is plotted for the fifth intermediate projection between planes $P2$ and $P3$. A more technical description of the Grand Tour can be found in Buja et al. (1996). Although the human eye is not well trained to recognize rotations in more than three dimensions, the Grand Tour can help to find structures like groups, gaps, and dependencies in the data.

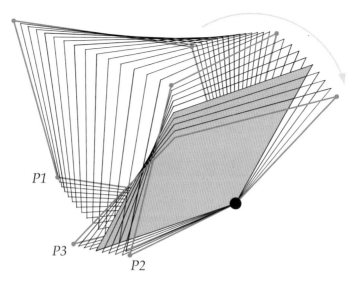

Fig. 2.13. *Sample path of a Grand Tour. The three base frames P1 to P3 are interpolated by intermediate projection planes to generate a smooth transition.*

Parallel Coordinates

Whereas scatterplots, 3-D rotating plots, and the Grand Tour all rely on orthogonal axes, parallel coordinates draw all variables in parallel, i.e. side by side. Every observation is then plotted for each axis/variable, and a connecting line is drawn for each observation between all the axes. This plotting technique was first published by Inselberg (1985). Figure 2.14 shows an example of a single observation x_i plotted in parallel coordinates.

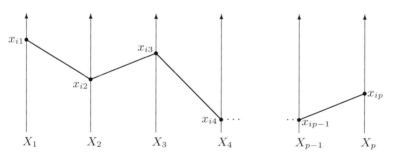

Fig. 2.14. *Observation $x_i = (x_{i1}, x_{i2}, x_{i3}, ..., x_{ip-1}, x_{ip})^T$ plotted in parallel coordinates.*

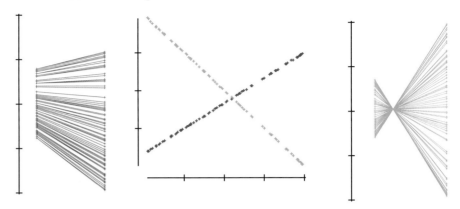

Fig. 2.15. *How lines in two dimensions translate into parallel coordinates.*

Parallel coordinate plots have several interesting geometrical properties that are more interesting for mathematicians than for statisticians.
The most important property in this sense is the point line duality, i.e.,
a line in two dimensions translates into a single point in parallel coordinates. Figure 2.15 shows how points on lines with positive and negative
slopes translate into parallel coordinates. In both cases, all lines intersect in a single point. If the slope is negative, the point lies between the
two parallel axes; if the slope is positive, the point lies outside the parallel axes. This property in two dimensions can be translated into higher
dimensions as well, to define hyperplanes.

Applications in visualization for data analysis look at features like
gaps, groups, and outliers within the multivariate display of a parallel
coordinate plot. Parallel coordinate plots are particularly good for identifying points that are outliers on individual variables and inspecting
their properties. Finding multivariate outliers that are not outliers on
any individual variables is much more difficult. Figure 2.16 gives four
examples of how clusters in two and three dimensions translate into features in parallel coordinates. All the clusters in Figure 2.16 can be seen
in one- or two-dimensional views, i.e., scatterplots. The advantage of parallel coordinate plots is to trace these features through many variables
simultaneously, which is almost impossible with lower dimensional plots.
Figure 2.17 shows five variables of the Cars2004 dataset. All five variables are highly correlated. Negative correlations — e.g., *mpg* and *engine
size* — show up in many crossings of the connecting lines. High positive
correlations show up in many almost parallel lines — e.g., *retail price* and
weight.

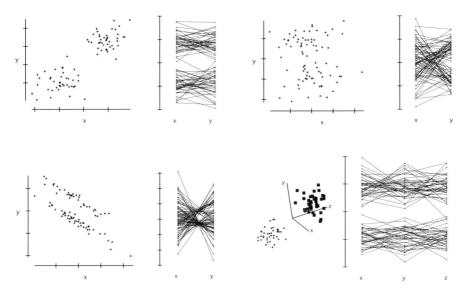

Fig. 2.16. *How groups can be identified in parallel coordinates.*

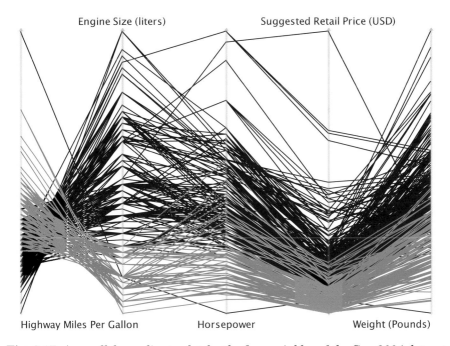

Fig. 2.17. *A parallel coordinate plot for the five variables of the Cars2004 dataset in Figure 2.12. All 4-cylinder cars are highlighted.*

2.4 Data on Mixed Scales

The last two sections dealt with either purely categorical or purely continuous data. When plotting data on mixed scales, trellis displays[2] have proven useful. Trellis Displays use a grid-like arrangement to place plots on the component panels. Each plot in a trellis display is conditioned upon at least one other variable. The biggest advantage of trellis displays is the common scale for all plot panels. This allows effective comparisons of the panel plots. A simple example of a trellis display can be seen in Figure 2.18.

Figure 2.18 shows a boxplot of *miles per gallon* by *number of cylinders*. Mileage can easily be compared between classes, because the scale does not change when visually traversing the different subjects. At this point, it is very important to note the difference between boxplots y by x and parallel boxplots. Whereas the latter plots a boxplot for **all** cases for every variable, boxplots y by x plot conditional boxplots for each group of the conditioning variable, i.e., each observation is only plotted **once**. Trellis displays are most powerful when up to two conditioning variables and more complex panel plots are used. The panel plot can be any arbitrary plot. Often fitted models are superimposed. Figure 2.19 shows an example of a trellis display for *Income* vs. *Working Hours* for the US Census dataset. The panel plots are conditioned for US Regions and Marital Status. Each scatterplot has a local regression smoother superimposed along with the local 95% confidence bands. The plot shows that *Income* grows with *Working Hours* and usually has a maximum at just over 60 hours. Highest incomes can be found for married people in the Northeast.

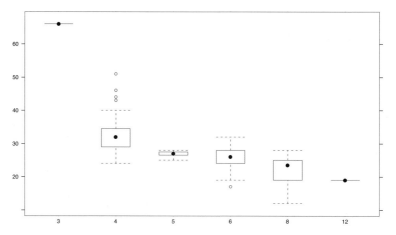

Fig. 2.18. *Boxplot* MPG *by* Cylinder *as simple form of a trellis display.*

___
[2] Trellis displays are called lattice graphics within the R package

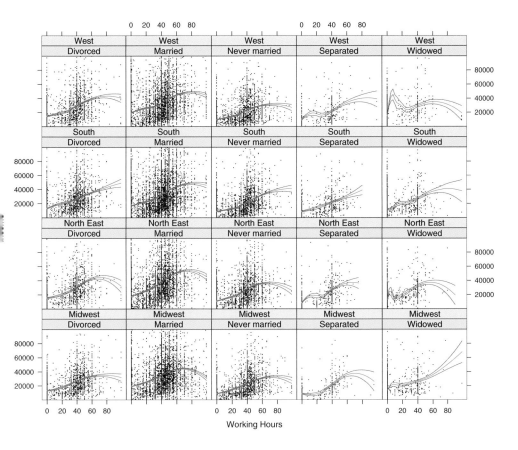

Fig. 2.19. *A trellis display showing scatterplots for* Income *vs.* Working Hours *conditioned by* Region *and* Marital Status *for the* Census *data. Each plot panel has a local regression smoother superimposed. Confidence bands have been added to illustrate the variability of the estimate.*

One problem with trellis displays is the fact that it is hard to judge the number of cases in a panel plot. For example, in Figure 2.19, the number of observations in the plot panels ranges from 328 to more than 10,000. When groups are smaller, it is essential to have indicators like confidence bands for scatterplot smoothers in order to judge the variability of an estimate, which is usually defined by the size of the underlying sample.

Another problem with trellis displays is the method of shingling. Shingling is used when a conditioning variable is not categorical but continuous. Shingling a continuous variable means to discretize the variable into a number of overlapping categories or intervals. If this is done automatically without any control and insight, the resulting categories can be hard to interpret and may even mislead.

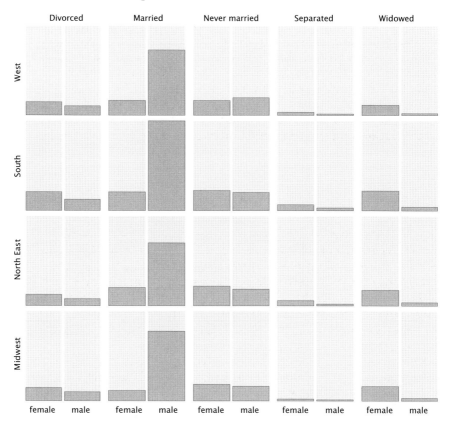

Fig. 2.20. *A multiple barchart plot for the variables* Marital Status, Region, *and* Gender *for the* US Census *data. The distribution of* Gender *clearly only depends on marital status and seems to be independent of region.*

Trellis displays and mosaic plots share the fact that data are plotted conditioned to a specific sequence of variables. Both plots use rectangular layouts to do so. Whereas mosaic plots are defined recursively within a two-dimensional rectangular region and can accommodate — at least theoretically — an arbitrary number of dimensions, trellis displays arrange plots in a hypercube-like layout. When looking at only two conditioning variables, the two plot types are quite similar. Typical modifications of mosaic plots use equal space for each cell. These modifications are same binsize, fluctuation diagrams, and multiple barcharts. These variations are described in Section 5.4.

Figure 2.20 shows an example of a modified mosaic plot for the variables *Marital Status*, *Region*, and *Gender* for the *Census* data. The panels are conditioned in the same way as in Figure 2.19, the "panel plot" is a barchart for gender. Obviously the similarity of trellis displays and mosaic plots is limited to the special case where there are two conditioning variables and categorical panel variables displayed in equally sized cells. This can be seen when adding a third variable to the examples in Figures 2.19 and 2.20. Adding a third conditioning variable to the trellis display will result in different pages for each level of the third conditioning variable. This is very similar to the way data cubes in OLAP (OnLine Analytic Processing) systems are arranged, respectively aggregated. In contrast, the mosaic plot is a recursively conditioned plot, i.e., a third variable will be introduced conditioned on the first two variables.

2.5 Maps

Maps form another special case in statistical graphics. Whenever there exists a geographical reference for a dataset, it is informative to be able to visualize the geographical reference together with the "non-geographical" parts of the data. Standard approaches display the statistical information on the map using glyphs or colours and are useful for presentation purposes. Interactive approaches linking geographic and statistical displays are better for exploratory purposes. Maps do not substantially abstract geographical information, as they usually use the same geometric representation of areas just on a smaller scale. Plotting geographical information opens up the whole area of cartography, which cannot be covered in this book. For a comprehensive overview of cartography issues refer to Kraak and Ormeling (1996). A wide array of geovisualization topics is covered in Dykes, MacEachren and Kraak (2005).

Maps with colour-shadings to represent quantities are called choropleth maps. Figure 2.21 gives four examples of choropleth maps. For each of the four plots, the median household income is plotted for all counties in the east of the United States. The upper left plot uses a linear gray scale for all values, i.e. the smallest value is assigned "white" and the largest value is assigned "black", and all other values are linearly interpolated between black and white. The same assignment method has been used for the upper right plot, except that the values now range from light yellow to dark red. The linear assignment has been changed to an equal probability scheme in the lower left plot, i.e. the colour shades have been assigned linearly to the ranks of the income values. The lower right plot shows additional highlighting for counties with a high proportion of young children. Obviously great care must be taken when using colours for plotting the map as well as for highlighting. A discussion of an interactive assignment of gray scale values in the context of mosaic plots can be found in Section 5.4.4. The use of choropleth maps as linked micromaps is discussed in Carr et al. (1998) and Carr et al. (2002).

For a discussion of the statistical exploration of geographical data, see Theus (2002a) and Dykes et al. (2005, Chap. 6).

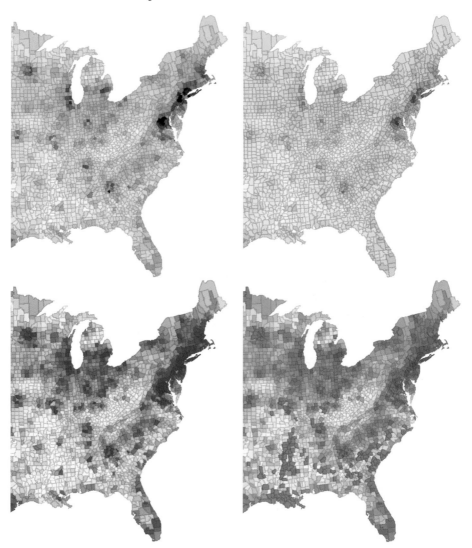

Fig. 2.21. *Four choropleth maps for the east-coast counties of the US, showing the median household income per county: (a) upper left: linear gray shading; (b) upper right: linear colour scale using values from light yellow to dark red; (c) lower left: equal probability colour scale; (d) lower right: equal probability scale with additional highlighting of areas with 8 or more percent of children under 5.*

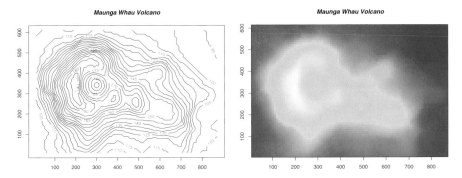

Fig. 2.22. *Left: A contour plot for the altitude at Maunga Whau volcano in Hawaii. Right: the same data shown in an image map.*

2.6 Contour Plots and Image Maps

Contour plots inherit features from both scatterplots and maps. Usually, three continuous variables are displayed in a contour plot. Two variables for the x-axis and the y-axis, and for the third variable z, contours are estimated and plotted. Estimating contours is not trivial and can lead to artifacts whenever extremes slopes occur in the data. In non-extreme cases, most algorithms give very similar and comparable results. An overview of contouring and surface displays can be found in the Appendix of Scott (1992).

The most natural use of contour plots is to actually map geographical altitude via a sample of altitude measures at given longitudes and latitudes. Figure 2.22 (left) shows an example of a contour plot for the Maunga Whau volcano in Hawaii. The different contour-lines[3], i.e., in this case lines depicting equal altitude, are annotated. Figure 2.22 (right) shows the same data in an image map. In an image map, areas in the same range share the same colour. Depending on the application, different colour ranges are used. The colour scheme used in Figure 2.22, from green, via yellow, and brown to white, is typical for plotting altitudes. Temperatures are usually depicted by a range from blue (cold) to red (hot). Thus, image maps are also often called heat maps. But contour plots also have an application closer to the statistics domain. They can be regarded as the two-dimensional generalization of a density estimator as described in Section 2.3.1 and Section 3.5. Obviously, a two-dimensional density estimator has to cope with the same problems as its one-dimensional counterpart. Features like gaps, or accumulations are hard to preserve for an estimator that tries to calculate a smooth result. If the data follow a non-pathological distribution, a contour plot can be a good approximation

---

[3] Contour-lines are often also called iso-lines, as values along these lines share the same value.

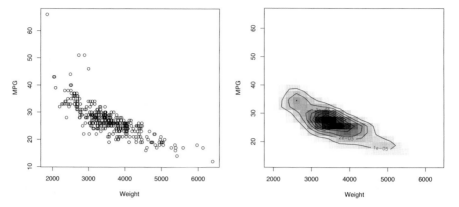

Fig. 2.23. *Left: A scatterplot of* MPG *vs.* Weight *of the Cars2004 dataset. Right: A contour plot for the density of the joint distribution of* Weight *and* MPG.

of the two-dimensional distribution. In Figure 2.23 (left), the variables *Weight* and *MPG* of the cars data are plotted in a scatterplot. The right plot in Figure 2.23 shows the density contours along with an image plot in gray scales. The outliers with high gas mileage are almost invisible in this colour mapping. Similar estimates can be generated using *tonal highlighting* and *α-blending* as discussed in Section 3.5. For a deeper and more elaborate discussion of density estimators in more than two dimensions, consult Scott (1992).

2.7 Time Series Plots

From a technical point of view, time series plots have a lot in common with scatterplots. In a time series plot, the x-axis is always set to be time, and the y-axis is the observed variable. Often it can be reasonably assumed that measurements could be taken at any arbitrary point of time, and so the points are connected by lines. Despite their similarity with scatterplots, time series plots are quite different in application. When looking at more than one time series, it is essential to overplot and rescale time series plots for efficient comparison. No matter how you look at time series data, the x-axis always represents time. This holds true for time series as well as for components of time series. Thus, many of the issues related to scatterplots, such as masking of structure through overplotting or density estimations, are meaningless for time series plots. Figure 2.24 shows a typical example of a time series plot. Four major stock indices are plotted for the 1990s. For a better comparison, all series are plotted in the same coordinate system, all aligned at the same time axis. Depending on the application, the scaling of the y-axis is modified. In the case of financial time series as in Figure 2.24, series are often normalized at 100% for

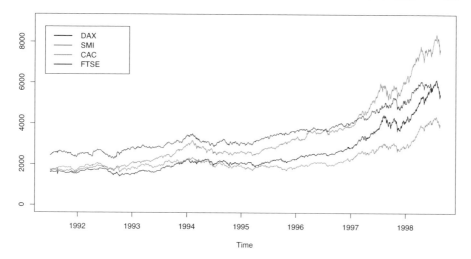

Fig. 2.24. *A time series plot for four major stock indices for the 1990s.*

a start of a particular time interval. Due to the many different and very specific applications in which time series plots are used, this topic is not discussed in this book. For further reading, refer to Wilkinson (2005) and Hochheiser and Shneiderman (2004).

2.8 Structure Plots

Structure plots depict properties of a whole dataset in a single plot. It is usually assumed that the dataset is a rectangular matrix with k columns and n rows — which is not necessarily true with complex datasets, but can often be achieved by transformations for at least a reasonable subset of the data. Depending on which structure is of interest, different kinds of structure plots can be used.

Tableplots

Tableplots try to capture the information of the complete dataset in a single plot. They use a spreadsheet-like layout of the data. Figure 2.25 shows an example of such a tableplot for the first 12 variables of the US Census dataset. In this plot, the data are presented in a table-like layout. Each column represents a variable, and each row is an aggregation of 250 cases. The data have been sorted according to *Age*. Each line shows the summary for the 250 aggregated values. For a continuous variable, the mean value is shown by a horizontal bar of corresponding size (cf. variable *Age*). For categorical variables, a horizontal bar is shown, which is subdivided according to the proportion of cases within the particular aggregation

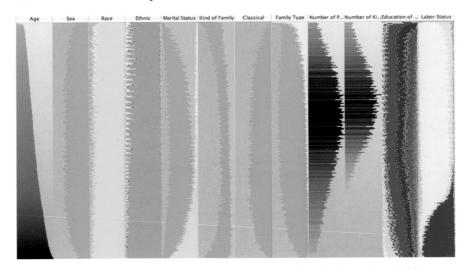

Fig. 2.25. *A tableplot of the US Census dataset for the first 12 variables.*

group (cf. variable *Sex*). Querying the graphics shows that the proportion of females is higher for both younger and older groups in the dataset. Almost all other variables in the plot show a strong interaction with *Age*, as for instance *Marital Status* does. Figure 2.25 was generated using the GAUGUIN software (http://www.rosuda.org/gauguin). Obviously, the aggregation mechanism guarantees that no matter how many cases are looked at, it is always possible to get an overview of all cases at once. Whereas only technical limits arise for the number of cases, the number of variables is still limited to the number of columns, which can be displayed — usually some dozens.

A similar approach can be found in the TableLens software (Rao and Card; 1994).

Missing Value Plot

Missing value plots focus on just a single attribute of the data: whether a value was recorded or not. Figure 2.26 (left) shows a missing value plot for ultrasound measurements and the birthweight of babies. In a missing value plot, a bar is drawn for each variable, which is divided into the proportion of missing and non-missing values. In the left-hand plot in Figure 2.26, we see that about 10% of the values of each of the ultrasound variables are missing and no cases have missings for the variable *Weight*. In the right-hand plot, the missings of the measurement *BIP* are selected. The highlighting shows that almost all of the missings in *BIP* are also missing in the other two measurements. Only a few values are missing at only one of the variables. There are no missings for *birthweight* itself.

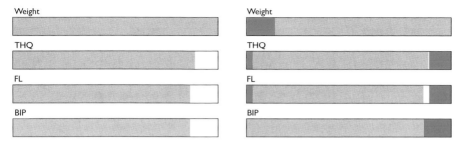

Fig. 2.26. *A missing value plot for the ultrasound dataset (left). All cases with missing values in* BIP *are selected (right).*

When missing value plots can be sorted according to the absolute number of missings in each variable, a check for monotone missingness can be easily performed. Stepwise selection of the missings from the smallest missing group to the biggest will show whether the selected missings are included in all successive variables. Monotone missingness is not only an interesting structural feature of the data but also makes a possible imputation of the missing values more effective.

From Figure 2.27 you can see that question 3 was the least popular and questions 2 and 4 the most popular. For most questions, the probability of passing the exam does not change too much, depending on whether a student did answer a question or not. Not so for question 7, where there is a probability of passing the exam of almost 70% for students who decided to take question 7 and a probability of only about 15% of passing for those who did not attempt the question. Question 6 shows the inverse interaction. This question was among the unpopular ones, and students who decided not to answer it had a higher probability of passing than those who did answer it.

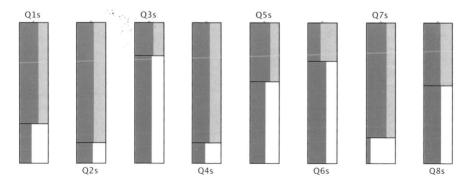

Fig. 2.27. *A missing value plot for the exams dataset. Students who passed the exam are highlighted.*

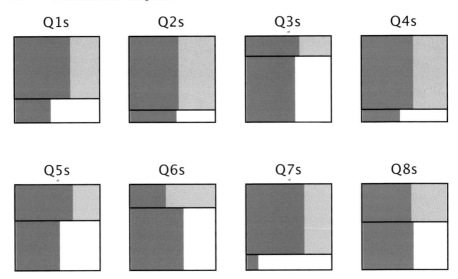

Fig. 2.28. *The same data as in Figure 2.27 with optimized aspect ratio.*

In this setting, it is easiest to judge the absolute sizes of the overall and highlighted groups. In Figure 2.27, the highlighting direction is chosen to be orthogonal to the primary discrimination between missing and non-missing. In doing so, it is possible to compare the highlighted proportions between missings and non-missings much more easily. To make the interaction between missing and non-missing and passed and failed better visible, the aspect ratio of the boxes can be adjusted to be 1 as shown in Figure 2.28. The importance of choosing the right layout and ordering of mosaic plots is discussed in Chapter 5.

3

Scaling Up Graphics

Martin Theus

This high man, aiming at a million.

Robert Browning, *A Grammarian's Funeral*

3.1 Introduction

This chapter investigates what influence the size of a dataset has on standard statistical graphics. Area plots like barcharts, histograms, and mosaic plots are relatively invariant against the increasing size of a dataset. Point plots, all plots that plot a glyph for each observation, have to cope with the problem of overplotting and an increasing number of extreme outliers. The definitions of some other plots, for instance boxplots, are examined to see if they still seem to be appropriate when datasets get really large. Problems, solutions, and modifications are presented, which are either based on more interactivity or on advanced plotting techniques.

3.2 Upscaling as a General Problem in Statistics

Jerome Friedman (2001) quoted his former boss Chuck Dickens at SLAC (Stanford Linear Acceleration Center), "Every time computing power increases by a factor of ten, we should totally rethink how and what we compute." Friedman applied the same idea to the size of datasets writing: "Every time the amount of data increases by a factor of 10, we should totally rethink how we analyze it." But unlike other disciplines, statistics is not very adaptive. It does not take long to find out that a χ^2-test will always give a significant result if only the sample is big enough, or vice versa, given a big enough dataset, there is no interaction between two categorical variables that is not significant. Unfortunately, most textbooks use examples of only a limited size. This upscaling problem can be found for many mathematical statistical tests, questioning the relevance of much of the statistical theory developed during the past century for problems that are not of small size. This problem has become more pressing recently, as the increasing sizes of datasets are linked to the increasing capacity of databases and to increases in computer power in general.

At first sight, statistical graphics might seem to be relatively robust to the scale of a dataset, but this is not true for many statistical graphics. Whereas area-based plots for categorical data scale up quite well, most statistical graphics for continuous data suffer badly when the number of plotted entities is large. For large datasets, these plots have to be modified or new plots have to be introduced.

Computer scientists working in the field of Data Mining — mostly coming from an information visualization perspective — have introduced new, specialized plots for visualizing massive datasets. Most statistical plots are not as specialized and are more generally applicable to a broader range of problems.

3.3 Area Plots

Area plots usually depict counts, i.e., a statistical summary of the data. Looking at a barchart, it is hard to find out whether the underlying sample was of size 100 or 10 million. The picture itself might look the same. The "only" thing that is affected is the interpretation of what can be seen. Small variations in the heights of the bars will be a weak hint for differences between the categories, if the sample is small. For a larger sample, small variations might be highly significant and clear evidence of a difference. Nonetheless, for large datasets almost any visible difference will be significant, regardless of its relevance.

Although histograms look different from barcharts (or they should, some users and some software packages mix up the two ideas very casually[1]), they serve a very similar purpose: depicting the counts of raw data. Barcharts can do this without any approximation, whereas histograms are approximations of the distribution of the data.

Mosaic plots break the barrier of dimensionality and depict several categorical variables at a time. The areas in mosaic plots are strictly proportional to the numbers of observations in the corresponding combinations of categories.

The next sections will take a closer look at the three major plots based on areas and show which problems arise with very large datasets and which additions might be needed to handle large datasets more effectively.

---

[1] Barcharts show counts for distinct classes of categorical data, whereas histograms depict a breakdown over a continuous interval of real valued numbers.

3.3.1 Histograms

Changing the anchorpoint and binwidth of a histogram can often uncover special structure in the dataset — gaps, thresholds, or discretizations. Because large datasets are often made up of data from several different sources, it is likely that different resolutions will be appropriate for different parts of the dataset. No one display is going to reveal all that might be found, and interactive controls are valuable for exploring different alternatives. It is necessary, however, to respect any structure in the data. The variable age from the US Census Data is only given in years and binwidths of 1.34 or 0.86 would both be meaningless and misleading.

Figure 3.1 shows a sequence of 9 histograms with binwidths of 2, 4, and 6 and anchorpoints 0, −1, and −2 for the distribution of *Age* in the *US Census* dataset. All the histograms here convey roughly the same information, though the dip due to the war years is not apparent in the displays on the right.

The histograms with binwidth 2 show an increase for the interval $[90, 92)$, which is due to rounding the age of all respondents of the study older than 90 down to 90. This artefact in the data can only be seen if the binwidth is set to be small enough and is a typical example of things one might find in a large dataset. Most rules to determine the number of bins in a histogram do not give sensible results for large datasets and cannot take account of any special structure (e.g., discretizations).

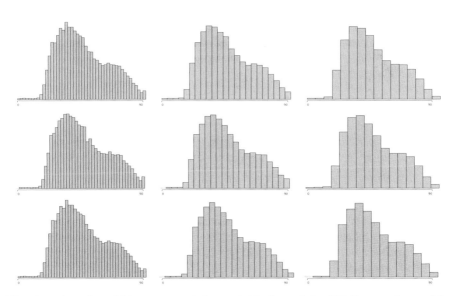

Fig. 3.1. *A series of histograms of the variable* Age *of the US Census data with binwidths 2, 4, and 6 (left to right) and anchorpoints 0, -1, and -2 (top to bottom).*

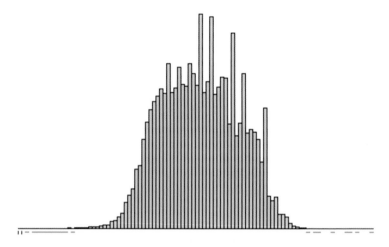

Fig. 3.2. *A histogram of logged amount of 450,000 bank deals. Small bars are invisible due to the big differences in cell counts.*

With large datasets, histograms also have to cope with very big differences in the bar counts, which often make small bars (almost) invisible. Figure 3.2 shows an example of a histogram of logged amounts of 450,000 bank deals. At both ends of the histogram there are bins with very small numbers of cases.

Special plotting techniques have to deal with this problem. One solution is to redmark the bins, as in Figure 3.2 (for redmarking see also Section 4.3.7). Another solution is to always assign bars a minimum size of a single pixel, which might lead to slightly incorrect binsizes, but ensures that information is not invisible.

3.3.2 Barcharts

In a barchart, the relative sizes of category bars do not depend on the sample size of the dataset. Consequently, it is hard to draw the right conclusions, not knowing the amount of data involved.

Approximate confidence intervals for bar heights, say based on a Poisson model, would give some guidance, but as datasets get bigger and bigger these intervals would become smaller and smaller till they finally degenerate and are no help at all. A 95% confidence interval for 100 cases represented by a bar of height 100 pixels would have a height of 40 pixels, and the corresponding interval for a bar of 100,000 cases would only take up 1 pixel.

In general, three kinds of categorical variables can be distinguished:

- **Fixed**

  The first class is the class of variables with a fixed number of categories no matter how big the dataset gets. For example, variables like *Gender, US State, Continent*, or *Blood-Group* have an a priori known number of categories.

- **A-priori Limits**

  The second class are discrete variables, where the number of categories to be observed is not known a priori, though limits can still be estimated. This class includes variables like *Make of Car* or *Native Country*.

- **No Known Limits**

  The third class is the class of variables with a number of categories, which grows with the number n of observed cases. This class includes variables like *Company Employed, Native City*, or *Favorite Beer*.

Obviously, the first two classes cause no special problem when datasets get very large, but the third does. Barcharts with dozens or even hundreds of categories need special interactive support to allow the user to navigate through the categories. These interactive functions comprise:

- Ordering
  - by Frequency
  - by Name
  - by Highlighting
    - · relative
    - · absolute
  - manually
- Grouping

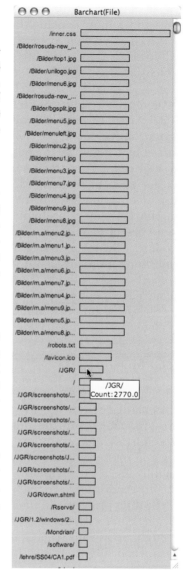

Fig. 3.3. *A barchart with more than 4,500 categories.*

These options will be discussed in more detail for barcharts and mosaic plots in Chapter 5 using the Internet Usage dataset.

To cope with many categories, a first step is to plot the bars horizontally rather than vertically. This permits printing most labels in full — which is important for a successful identification of the bars. A second step is to allow scrolling in barcharts. Figure 3.3 shows part of a barchart of the number of hits of files of a web server. This barchart has well over 4,500 categories, which can be easily investigated in a horizontal bars layout, using sorting options and scrolling. The barchart was reversely sorted by frequency. In such a situation, it is almost impossible to find a specific category without sorting options.

For even more categories ($> 1,000$), Eick and Karr (2002) show an implementation of logical zooming in barcharts. In a global view of the barchart, the bars are plotted with a minimal width of a single pixel. If there are more bars than pixels, adjacent bars must be joined to accommodate all categories in a single view. Zooming in widens the bars more and more until the usual barchart view appears again. This technique is most effective when the bars are sorted according to absolute size to achieve a structured plot of the information.

Section 4.4.3 presents further details on zooming issues for a variety of plots.

3.3.3 Mosaic Plots

Mosaic plots display categories or crossings of categories of categorical data using tiles rather than bars. The areas of the tiles of a mosaic plot are proportional to the numbers of cases in the corresponding groups. Barcharts with very many categories can be made scrollable and sorting and regrouping can support more detailed investigation. Mosaic plots usually use a fixed window size, which is then subdivided recursively into the different tiles. Sorting, grouping, and zooming are all important.

Scaling problems in mosaic plots arise as the number of tiles becomes larger. Plots with about 50 to 100 tiles are usually still easy to interpret, though, as always, this depends on how much structure there is in the data. Basically there are two reasons why the number of tiles becomes large. Neither is specific to large datasets, but both are more common with large datasets.

The first reason arises when the variables being combined have many categories. As long as there are only two variables, the matrix-like structure of the mosaic plot is not difficult to read, even if there are dozens of categories. Adding a third or fourth variable will tend to overload the plot.

The second reason is due to including many variables. The number of tiles grows exponentially with the number of variables in the plot. As known from high-dimensional data analysis, points become more and

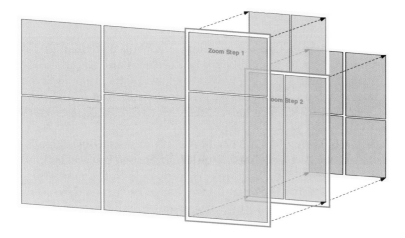

Fig. 3.4. *Logical zoom example in a mosaic plot.*

more sparse the more dimensions are considered. With categorical data this means that more and more empty intersections of categories are observed. Even in a very large census dataset, no observations are expected to fall in the category of females with a doctoral degree, below the age of 20, having 3 or more children, and earning less than $20,000 a year. Interestingly, large numbers of empty cells may be beneficial, if clusters of data stand out better.

There are two options for dealing with very many tiles in a mosaic plot. Either the plot is made scrollable or zooming functionality is added. Both options need additional navigational help. In a scrollable mosaic plot, it would not be possible to get a full view of the mosaic plot as a whole. When zooming into a mosaic plot, the user is likely to lose the context of where he or she was focusing on. Both solutions can benefit from an overview window or bird's eye view to give orientation as long as the zoom factor does not get too large.

This leads to the concept of logical zooming in mosaic plots. To avoid clutter in a mosaic plot, a maximum number of cells to be plotted can be defined. The recursion stops if the number of cells would exceed this maximum. Thus, not all variables might be included in the plot. The user then may select a category in the first variable and zoom in on this category. In the zoomed view, the next variable, which was not included at the end of the recursion due to the stopping criterion, is then included. This can be repeated until all variables are included. A separate overview window, containing only the conditioned variables, shows the current zoom location to guide the user.

Figure 3.4 shows a simple example of how logical zooming in mosaic plots looks. (For logical zooming, see also Section 4.4.3.)

3.4 Point Plots

Point plots all suffer from the problem of trying to depict every single observation with its own glyph.[2] For scatterplots and parallel coordinate plots, this is obvious. Although boxplots do not plot a glyph for every case, the number of glyphs to plot is still a linear function of the number of cases n, because outliers are individually plotted, as discussed in the next section.

Plotting too many glyphs usually results in meaningless black areas, due to overplotting. This effect is quite similar to an overloaded signal that is then clipped. The black area corresponds to the maximum signal level. To avoid clipping, the overall level has to be turned down. This principle can be applied to plots by using tonal highlighting or α-blending. Both methods reduce the weight of a single observation and offer a rudimentary density estimation of the data. These methods will be introduced in Section 3.5.

3.4.1 Boxplots

Being based on robust data summaries (i.e., median, upper, and lower hinges, and upper and lower inner and outer fences), boxplots might be assumed to be insensitive to increasing dataset sizes. This holds true for the summary statistics, but not for the outliers. A case x_i is an outlier in a boxplot when

$$x_i < x_{0.25} - 1.5 \times (x_{0.75} - x_{0.25}) \tag{3.1}$$

or

$$x_i > x_{0.75} + 1.5 \times (x_{0.75} - x_{0.25}). \tag{3.2}$$

(Note that the 0.25- and 0.75-quantiles are used here instead of upper and lower hinges, which makes a negligible difference when the dataset is large.) Assuming a standard normal distribution gives an IQR (Interquartile Range, i.e., $x_{0.75} - x_{0.25}$) of 1.348. Thus, values outside the interval $[-2.698, 2.698]$, about 0.69% of the sample, are classified as outliers. With this result, about 7,000 outliers would be expected for a dataset of size 1 million in a single boxplot. Obviously it would not be possible to investigate every single one of them. Looking at a standard log-normal distribution, which seems to be more appropriate for values that are determined to be bigger than zero, the upper fence is about 4.14, leading to almost 8% outliers. Figure 3.5 shows the histogram of a sample from an $N(0, 1)$ standard normal distribution of size 100,000 along with a boxplot of the same data. All outliers in the boxplot are highlighted and marked in the histogram. By definition, any point that is far enough in the tails of the

---

[2] A glyph — which is best translated as some "small graphics symbol" — can be as simple as a single point.

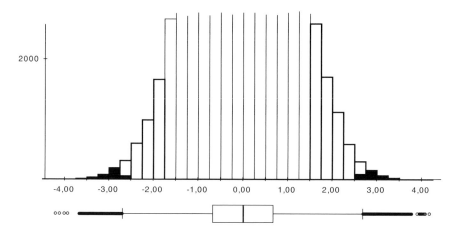

Fig. 3.5. *All 763 outliers highlighted in a sample of a standard normal distribution of size 100,000.*

distribution will be classified as an outlier. In practice, only extreme outliers or distinct outlier clusters (for instance, where there is a gap to the rest of the data) will be of immediate interest. Linking the possible outlier cases to their values on other variables may uncover unusual multivariate cases. Section 3.6 will investigate possible modifications more closely.

3.4.2 Scatterplots

Scatterplots do not rely on summaries of the data but on the raw data themselves, so they are most likely to deteriorate when datasets get very large. Some simple calculations are helpful. How many pixels does a window of average size on a computer screen hold? A typical window of size 800×600 holds almost 500,000 pixels, so if the data were distributed more or less uniformly over x and y, about 100,000 points could be placed in this window without substantial overplotting, provided that the data do not exhibit discretization. Overplotting is a problem only when data accumulate at certain points or are measured at a fixed resolution. Note that plotting data on a computer screen already discretizes the data to the pixel resolution of the screen.

Figure 3.6 shows a scatterplot of a random sample of 100,000 points from a bivariate distribution of independent standard normal $N(0,1)$ and standard uniform $U[0,1]$ distributions. Overplotting is not a real problem here, and any departure from the sampled structure would be readily seen.

Of course, real-life problems will show overplotting and the major aim is then to get an idea of the two-dimensional density. Many suggestions have been made for two-dimensional density estimation. In the context

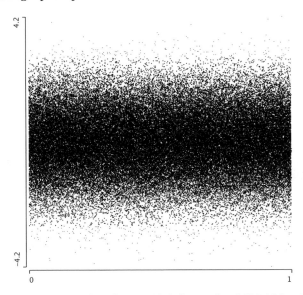

Fig. 3.6. *A scatterplot of an artificial sample of 100,000 points.*

of large datasets, the efficiency of an algorithm plays the most important role. Because large datasets usually give a good approximation to some underlying distribution (with large datasets often a mixture of several underlying distributions), the optimality of an algorithm is less important than its efficiency.

Section 3.5 will investigate density estimation techniques in the light of large datasets.

The artificial data in Figure 3.6 do not have any extreme outliers or special structure. Real datasets often have extreme values and structure hidden in specific subsets. This structure cannot be revealed in a global view of all the data. Zooming functions can help to focus on special subgroups.

Zooming should be implemented hierarchically, i.e., not only predefined fixed zoom steps should be offered. In a hierarchical zoom, the "zoom-in" area may be defined arbitrarily. A "zoom-out" should step back to the last zoom level. This technique allows a structured search deep into very detailed structures of a scatterplot without losing context.

Figure 3.7 gives an example of how to visualize micro structure in a scatterplot. Whereas the plot on the left gives little insight into the specific structure of the two-dimensional data, the plot on the right shows some structure invisible in the default view. It was generated by zooming in by a factor of 2 on both axes and applying α-blending. The plot shows an income threshold at about \$100,000, which is also visible as a linear threshold for *Household Income*. Some accumulations at the full ten-thousand amounts are also visible.

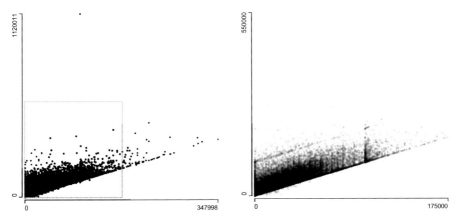

Fig. 3.7. *Standard scatterplot of Total Household Income vs. Income of Household Head (left), a more detailed zoom with α-blending (right). (Zoom region indicated in red.)*

3.4.3 Parallel Coordinates

Parallel coordinate plots differ substantially from scatterplots. Whereas in scatterplots just two dimensions are plotted on a two-dimensional area, parallel coordinate plots only use one dimension per variable and display up to 10, 20, or even more variables in parallel. Heavy overplotting is expected on the individual axes and even more between the axes. Connecting the points along each variable with lines makes the plot look like a black band of lines if more than a few thousand cases are plotted. It is difficult to recognise any structure, such as groups or patterns, any more.

Figure 3.8 shows a parallel coordinate plot of 572 observations of Italian olive oils. The parallel coordinate plot shows the content of 8 fatty acids. Although there are fewer than 600 cases, the plot is essentially a black band, not revealing any structure within the band.

As with scatterplots, the density of lines becomes more and more important as the number of cases increases. But whereas with scatterplots very good results can be achieved using tonal highlighting, the much greater amount of overplotting in parallel coordinates limits the levels of data that can be plotted more strongly. Although it is possible to reduce the weight of a single line (i.e., the α-values), detailed information will be lost as well. In this situation, non-linear functions, which map an overplotting density to a saturation of the colour, may be tried.

Figure 3.9 shows the same data as plotted in Figure 3.8. By applying α-blending, the group structure of several of the 9 Regions is now visible.

Zooming can also help to declutter parallel coordinates. Zooming in parallel coordinate plots means to individually zoom in on a single axis of the plot. Many of the connecting lines from adjacent axes may then

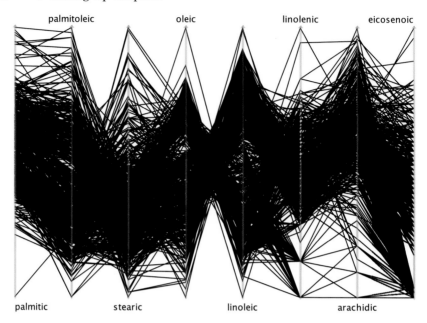

Fig. 3.8. *A parallel coordinate plot of 572 measurements on Italian olive oils. The heavy overplotting obscures the group structure in the data.*

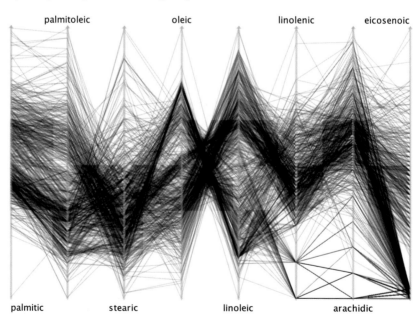

Fig. 3.9. *The same data as used in Figure 3.8 with α-transparency. Several of the 9 Regions can be seen.*

connect to points that are clipped. Successive zoom operations can peel into a region of high density step by step.

3.5 From Areas to Points and Back

Just as a histogram is a crude approximation for the density of a single variable, binning (cf. Carr et al.; 1987) is an approximation for the density of the joint distribution of two variables (see also the definition on page 129). Results regarding choice of histogram anchorpoints and binwidths to minimize approximation errors can usually be generalized to the two-dimensional case. Whereas there are no degrees of freedom of how to distribute the points over the intervals in the one-dimensional case, there is one extra choice to be made in the two-dimensional case: the shape of the bin must be specified. Only triangular, quadratic, or hexagonal bins qualify as bins, because all bins must have the same shape and cover the whole region. Although one would expect noticeable differences in the quality of the approximation, Scott (1992) showed that the differences are only marginal. Thus, quadratic binning can be used on the grounds of simplicity and efficiency.

In two dimensions, adjacency becomes more interesting. Besides *direct binning*, where all points are assigned with weight 1 to the appropriate bin, *linear binning* has been proposed. When binning linearly, all adjacent bins get a weight according to the inverse of the distance from a point to the center of the bin. All weights for a single point must sum to 1. Linear binning is computationally very intensive and thus not really relevant for large datasets.

There are two reasons why switching from scatterplots of the raw data to binned data displays is a good thing. The first reason can be that the dataset is too large to fit in computer memory. In these cases, or cases where data must be handled inside databases, the data have to be retrieved via a database query. Queries that summarize data may last some seconds, but the amount of data that is returned is usually far less than retrieving the complete variables and can easily be handled by the software. The other case where raw scatterplots are no longer workable occurs when the number of points cannot be rendered fast enough on the screen. In this case, the approximation via binning can be displayed much faster.

Most optimality criteria are no longer relevant when a dataset gets large. Using finer and finer grids, the variation certainly grows. When the binsize is down to one pixel, the binned plot is equivalent to a scatterplot. Whereas the number of points to plot in a standard scatterplot is of order n, the number of bins to plot is limited to the number of pixels in the plotting area and thus does not depend on the size of the dataset. Binned plots can ideally be used to give an approximation of a scatterplot.

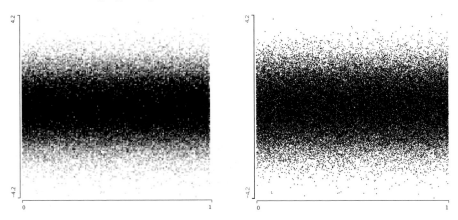

Fig. 3.10. *The sample data from Figure 3.6. Raw scatterplot (left) and binned to a 256 × 256 binning grid (right). Both displays render the structure of the data sufficiently well.*

Figure 3.10 shows the data from Figure 3.6 both as a raw scatterplot (left) and as a binned scatterplot. The binning grid is 256×256, which is about one-quarter of the window resolution. The overall impression of the two-dimensional distribution does not differ from the one in Figure 3.6.

When zooming into the plot, a logical zooming may switch to a display of the raw data, as soon as the number of points to retrieve and plot is small enough to be handled gracefully.

Figure 3.11 gives an example of logical zooming in scatterplots. The figure shows data for bank deals, plotting trade profit against amount. The first three plots (left to right and top to bottom) show binned data on a grid of 150×150 bins. The two first zoom steps (from left to right) each zoom by a factor 25^2, the third step zooms by a factor of 10^2. In this example, the self-similarity in the data is striking. Querying this plot shows the bin intervals and the corresponding counts. Whereas most bins include only a couple of points, the bins close to the origin still cover thousands of points. The plot itself displays about 190,000 points. The third zooming focuses on an area that does not include the origin and thus cuts down the number of points to plot more drastically to about 5,000. Because these 5,000 points can easily be handled in a scatterplot, they are now plotted as individual points.

Another use of binning methods is to estimate two-dimensional densities of pairs of variables. Methods like ASHs (average shifted histograms) and kernel-density estimations using various kernels have been proposed, as in the one-dimensional case.

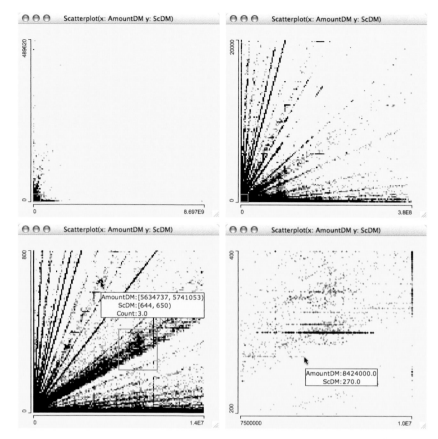

Fig. 3.11. *Zooming into 300,000 points. Except for the plot on the lower right, all plots use binning to approximate the scatterplot. The lower right plot shows 4,500 points in the zoomed area, which is only $(25 \times 25 \times 10)^{-2} = 6250^{-2} = 3.9 \cdot 10^{-7}$ the area of the plot top left. The queries in the lower row show different results according to the plot type (binned or classical). (Scales have been transformed for reasons of confidentiality; zoom regions are indicated in red.)*

3.5.1 α-Blending and Tonal Highlighting

When the α-channel of a graphics system is used for α-blending, a kernel-density estimator using a rectangular kernel function (and thus a symmetric kernel) is, in effect, obtained. The α-channel specifies the transparency of the colour of an object. The α-channel can be specified as a fourth channel besides the three RGB channels to specify a colour. An α-value of 0 corresponds to complete transparency and a value of 1 (or 100%) to complete opacity.

Fig. 3.12. *Four gray colour plates with an α-value of 0.25 (left). Using colour makes the interpretation of the resulting colours harder (middle). A zoomed detail of a scatterplot using α-blending (right).*

Figure 3.12 gives an example of how colours with α-transparency applied blend with their background. A further discussion of the interaction of highlighting and α-blending can be found in Section 4.4.2.

Whereas α-blending relies on the graphics system to "add" overplotting points, tonal highlighting counts the overplotting information for each pixel/bin explicitly. The mapping from counts to gray levels is done via a (possibly linear) transfer function.

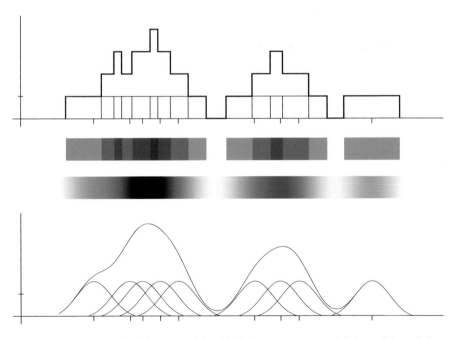

Fig. 3.13. *An example of how tonal highlighting compares with kernel-based density estimators.*

Theus (1996) showed that tonal highlighting is equivalent to kernel-density estimation as long as the kernel function k is symmetric, i.e.,

$$\hat{f}_{\text{th}} = \frac{1}{cn} \sum_{i=1}^{n} k\left(\frac{x - x_i}{c}\right) = \hat{f}_{\text{kernel}} = \frac{1}{cn} \sum_{i=1}^{n} k\left(\frac{x_i - x}{c}\right) \quad \text{for} \quad k(x) = k(-x).$$

$$(3.3)$$

for all kernel functions k

$$\int_{-\infty}^{\infty} k(x)dx = 1, \quad \int_{-\infty}^{\infty} k^2(x)dx < \infty, \quad \left|\frac{k(x)}{x}\right| \to 0 \quad \text{for} \quad |x| \to \infty. \quad (3.4)$$

Figure 3.13 illustrates the use of tonal highlighting for a crude density estimator. For 9 sample points, the resulting one-dimensional density estimates via tonal highlighting (upper plot and gray scale below) and via a normal kernel-density estimator (lower plot and gray scale above) are displayed. Obviously the result from the kernel-density estimator — which uses a normal kernel — is much smoother. This will be less relevant if the number of points is larger. On the other hand, features like gaps or other artefacts are more likely to be smoothed out by a kernel-density estimator that has a kernel function that is too wide.

3.6 Modifying Plots

Modifying the way graphs are plotted does not affect the essential definition of a graph. Making barcharts more interactive or using tonal highlighting or α-blending in scatterplots does not change the basic definition: bars are still bars and points are still points, even if their brightness may vary.

For boxplots, the number of outliers is likely to grow as a linear function of dataset size n, cf. Section 3.4.1. As long as datasets are small, Tukey's definition is faithful towards the way outliers would be identified and perceived. But for larger datasets, the definition is not so useful. The idea of plotting outliers in a boxplot was to identify extreme points in a sample, which might be worth investigating individually more closely. This is feasible as long as the number of outliers is not larger than a couple of dozen points.

Looking at Figure 3.5, points that look like outliers can be seen even in a boxplot of 100,000 points. Four points at the lower end and 10 points at the upper end clearly seem to be "further out" than the rest of the sample.

An updated definition of outliers in a boxplot should obviously take the density into account, to avoid plotting hundreds of points melting into a line lengthening the whiskers without allowing an individual investigation of the points. Furthermore, there should be a sensible gap between the whiskers and the first point classified as an outlier.

Even if a general definition does not exist for all large datasets, any solution that enables the identification and selection of outliers is acceptable, because it makes boxplots workable again.

3.7 Summary

The design and implementation of statistical graphics should pay attention to the challenges from big datasets. For many users, this has not been an issue up till now and so some statistical and graphics packages can have problems with graphics of more than 10,000 cases.

However, most of the plots used in statistical graphics can be scaled up to be usable with large datasets. Areal plots for categorical data are quite robust against large data glyph-based plots do have more serious problems. Modifications like α-blending or binning, interactions like (logical) zooming and panning, or interactive reordering and grouping are of great assistance when dealing with large datasets.

In general, all statistical graphics that summarize the data, and plot some version of these summaries, will scale up to large datasets. Barcharts, for instance, plot the breakdown of a categorical variable, which is a sufficient summary to fully describe the data. Binned scatterplots show an approximation of the underlying scatterplot and have a complexity that depends on the (constant) size of the binning grid rather than on the size of the dataset.

4

Interacting with Graphics

Antony Unwin

What is your substance, whereof are you made,
That millions of strange shadows on you tend?

William Shakespeare, *Sonnet LIII*

4.1 Introduction

This chapter examines how interaction can extend the power and potential of graphics to deal with large datasets. The first part of the chapter looks at interactive graphics in general, and the second part concentrates on applying these ideas to large datasets. Large datasets are more difficult to visualize than small ones for the reasons outlined in Chapter 1. Interactive graphics offer effective methods for surmounting these difficulties. A static graphic is fixed and cannot be adjusted or extended. An interactive graphic can be queried and altered to supply more information.

In general, increased dataset size leads to increased complexity and requires more sophisticated tools, whether for querying, for scaling, for linking, for zooming, or for sorting. This may not be apparent when a large dataset is first examined, as interaction can be very effective in revealing what information is there on the surface. It is when you want to go to deeper levels and gather more detailed information that the need to extend basic interactive tools becomes clear.

Besides extending and revising existing interactive tools, new interactive ideas can be introduced to deal with the additional problems caused by the sheer size of large datasets. Subsetting, aggregating, transforming, and weighting are obvious targets in this respect, but so is the task of just managing data analyses.

4.2 Interaction

Interaction can be introduced convincingly and attractively by selecting a case in a scatterplot of a small dataset. Everyone can see the benefits of being able to get the exact values of a point's coordinates and being able to link a case to displays of other information.

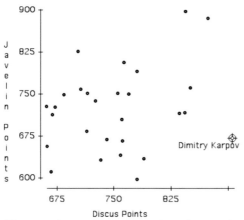

Figure 4.1 shows data from the Decathlon event at the 2004 Olympics in Athens. Dimitry Karpov got the highest number of points for the discus, but was one of the weaker performers in the javelin.

With large datasets, things are rather different. Very few points may be individually queried, and while linking from a group of points provides information about the group as a whole, this is not as interpretable as for a single point. Of course, interaction is much more than just querying. There

Fig. 4.1. A simple query to identify the athlete who has the highest score in the discus and a low score in the javelin.

is linking and the various ways of changing a plot's characteristics, like rescaling, zooming and sorting.

The term "statistical objects" is important. Graphic displays are built up of many individual components. For instance, a bar in a barchart has a border and an interior. Neither of these are statistical objects, only the bar as a whole. It may be useful to be able to interact with the individual components to vary the appearance of a display, but that is not the kind of interaction meant here. Similarly, an axis as a whole is a statistical object, though it is made up of several separate graphic components. Working interactively means that statistical objects on screen can be directly manipulated without the user having to resort to menus, dialog boxes, or command-lines. This is a fairly strict definition of interaction, sometimes different degrees of interaction are relevant for actions that are less frequently used, but it allows us to concentrate on the essentials of the concept.

The need for graphics in analysing large datasets is just as great as in analysing small datasets. Graphics are excellent for identifying "obvious" results (e.g., comparing two parallel boxplots that do not overlap), but not for detecting more "statistical" differences (e.g., whether the means of two overlapping subgroups differ). On the other hand, everything tends to be highly statistically significant with large datasets, and graphics are valuable for directing attention to differences that have some practical relevance. As with small datasets, interaction is helpful in large datasets for exploring first impressions to assess what validity they may have — by querying the values, by linking to other displays, by rescaling axes, by sorting data, or by zooming in. Without interaction, it is difficult to evaluate ideas gained from graphics.

4.3 Interaction and Data Displays

Pinning down the essentials of interaction for data displays is tricky. The problem has partly to do with people's notions of interaction itself. As Klinke and Swayne found when they edited a special edition of *Computational Statistics* in 1999 on interactive graphical data analysis: "There is some disagreement about the meaning of the term 'interactive graphics'." The word interaction has been used in a variety of ways in computing over the years and some of the older, weaker uses have survived longer than might have been expected. The main reason, however, for any difficulty in developing a theory of interactive graphics is that new interactive possibilities are continually emerging, causing rethinks of earlier ideas.

Both Theus (1996) and Unwin (1999) have proposed possible structures. Integrating the two suggests that there are three broad components of interaction for statistical graphics:

- Querying
- Selection and linking
- Varying plot characteristics.

Querying is a natural first activity to find out more about the display that you have in front of you. Next, you select cases of interest and link to other displays to gain further information. At this stage, it's helpful to vary the form and format of displays to put the information provided in the best possible light.

4.3.1 Querying

Querying enables you to find out detailed information about features that grab your attention. What is the value of that outlier? How many missing values are there? How big is the difference in size between two groups? The ability to browse displays quickly and without fuss is very useful for checking first impressions. Early interactive querying was restricted to returning the (x, y) coordinates of a point, usually at some distant output line. Now, it is not only possible to deliver the information directly at the location of the object queried, but much more sophisticated information can be returned as well. The key factor is that querying of any statistical object — be it a point, a bar, an axis, a variable or even a whole display — should always follow a consistent scheme. Making the display of the results of a query temporary avoids screen clutter and dispenses with the need for additional commands to remove information.

Providing different levels of querying is an elegant way of aiding the analyst in an unobtrusive manner. For instance, a standard query of a pixel in a scatterplot reports how many points are displayed there (if any), how many of them are highlighted and what range of X and Y values they represent. An extended query provides information on the values of other variables for those points. In Figure 4.1, you could also query where Karpov finished overall (he won the bronze medal, 168 points behind the winner, Sebrie, somewhat less than the difference between them in the Javelin). For larger datasets it would make sense to offer a further level to provide local distributional information, including mini graphical displays. It is tempting to consider multiple levels of querying, but this would counteract the desired simplicity of interactive methods. Three information levels — default, extended, and deep — should suffice. Deciding what these levels should offer for different graphical displays will involve contentious but constructive discussions.

For an example of queries prompting different levels of detail and complexity, consider again the example of Figure 4.1. Figure 4.2 shows the same scatterplot as Figure 4.1. The leftmost plot shows a simple query that is not object-related, but just gives the location of the mouse pointer in the coordinate system. The middle plot is the standard query of an object in the scatterplot — in this case the results of the winner of the Athens 2004 decathlon — which returns the exact values (x_i, y_i) of the point queried. The plot on the right shows an extended query, which gives additional information, not only related to the variables used in the plot, but to other information about this object. This is the simplest form of an extended query, but there could be other versions too, linking to further external information.

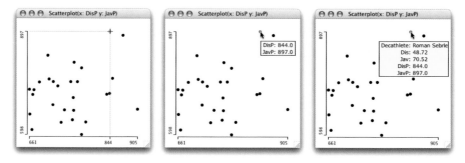

Fig. 4.2. *Example of queries at different levels of detail.*

4.3.2 Selection and Linking

Selection is used for identifying a group of cases of interest, using linking
to provide more information about them in other displays, commonly using highlighting of some kind. Highlighting means that linked cases are
displayed differently from others. (Querying is actually a special form of
selection and linking, which does not use highlighting but presents information for selected objects temporarily.) Typical applications would be
to investigate the properties of a cluster of points in a scatterplot or to
look at the patterns of values for a subset of the data. In small datasets,
that is about all you need and selection is a straightforward process. In
Figure 4.3, the oils from Inland-Sardinia from the olive oil dataset have
been selected. The scatterplot shows that all have very similar linoleic
and oleic values.

A more sophisticated form of selection is to use brushing: while the selection tool is moved across the data in a display, the selection is updated
in real time. This is an interesting option for exploring conditional structures, but the continually changing linked graphics are more difficult to
understand than with a simple selection and stable linking.

Linking is only possible if multiple views of the same data can be
displayed simultaneously. Software packages, which make that awkward
rather than intuitive, inhibit the use of linking. It is essential to think of
many displays contributing to an overall picture and not to aim for some
"optimal" single display. In the *Bowling Alone* dataset, there are almost
400 variables, and exploring groups of them together can be very informative. Figure 4.4 shows a barchart of the variable Fistfight ("I would do
better than average in a fistfight") with the options "Generally/Definitely

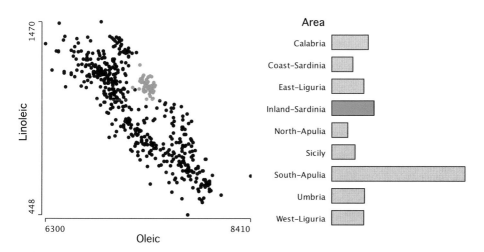

Fig. 4.3. *Simple linking between two plots. A bar in the barchart has been selected,
and those cases are highlighted in the scatterplot.*

agree" selected. The plots would be larger on screen and easy to read. Spineplots of eight possibly related variables have also been drawn and aligned to use space efficiently. The figure here is just to give an idea of what is possible for exploratory analysis, it is not intended to be of presentation quality. Some associations are immediately obvious:

- men have a higher rate (lowest middle display);
- as do those who believe there should be a gun in every home (lowest left display).

Some are non-existent:

- citysize seems irrelevant (middle top display).

Some are worth further investigation:

- number of children at home (right top display);
- number of children in education (right middle display).

Of course, these are only bivariate associations and there are doubtless multivariate structures here. Furthermore, information on age is not included and that is likely to be important. Nevertheless, the benefits of being able to display several graphics together are obvious. It is more likely to be done, if they can be drawn easily and flexibly. This is one of those supportive organisational tasks that software could be designed to carry out for us but usually does not.

4.3.3 Selection Sequences

Selections involving several variables can be made with selection sequences (Theus et al.; 1998). The idea is to combine the intuitive approach of interactive selection with the flexible power of editable code. All selections are made by clicking or grabbing the relevant parts of graphic displays, using the appropriate selection modes chosen from a tools palette or pop-up menu. The individual components of the sequence are automatically stored, so that they can be graphically altered or amended without affecting the rest of the sequence. DataDesk (Velleman; 1997) has offered an interactive selection capability for many years, but the individual components of the selection could not be changed without resetting the whole sequence. Selections written in code, for instance in R or in SQL, could be edited, but the process is not very intuitive. Both MANET (Unwin et al.; 1996) and Mondrian (Theus; 2002b) incorporate selection sequences. A similar approach was implemented by Hochheiser and Shneiderman (2004) in the TimeSearcher tool for analysing time series.

As an example, consider selecting the group of males with two children at home who believe there should be a gun in every home. Click on males in the barchart of gender, switch to intersection mode, select the

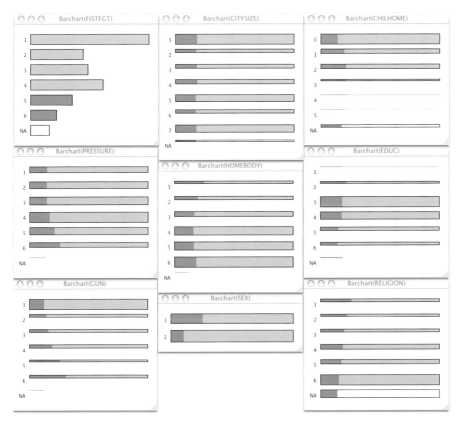

Fig. 4.4. *Nine related, linked variables out of just under 400 variables in the Bowling Alone dataset. (For presentation purposes on the printed page, this graphic has been drawn smaller than it would appear on screen.)*

bar for two children in the family barchart, and finally select the relevant bars in the gun barchart. Because this combined selection is equivalent to drilling down, the order of selection is not important here. The only thing that matters is that the first selection is a normal selection and that the following two are intersection selections. More complicated combinations are possible involving other selection modes where order is important, but drilling down is by far the most common selection sequence used. Editing a selection sequence is easy. If you want to look at the group of males with no children who believe there should be a gun in every home, you just need to click on the bar representing those with no children in the family barchart. The mode of that selection component and the rest of the sequence remain unchanged.

A selection sequence provides information on a selected subgroup, but no information on other complementary groups. This means that no comparisons can be made. It is only through making comparisons that statistics can be interpreted. Considering the survival rate of all females in first class on the Titanic, it would be appropriate to compare it with the rate for males in first class and with the rate for females in other classes (but not with *all* other passengers and crew). These comparisons are possible in mosaic plots by choosing the relevant ordering of variables, where each comparison requires a separate ordering. Note that these are exploratory analyses. If it was known in advance what selections were of most interest, they could be pre-programmed together with the necessary groups for comparison.

One solution would be to automatically produce a set of complementary mosaic plots for the selected group by each of the defining variables

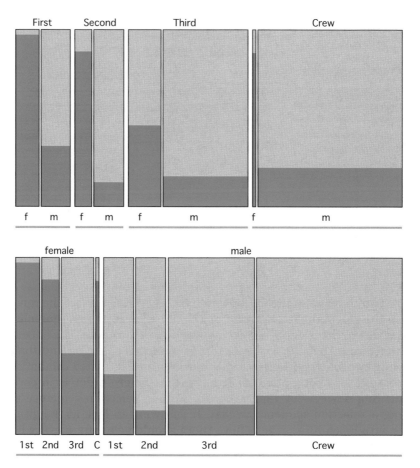

Fig. 4.5. *Doubledecker plots of the Titanic dataset with survivors highlighted. Above: Class by Gender. Below: Gender by Class.*

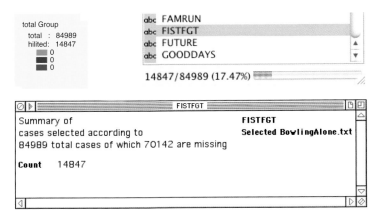

Fig. 4.6. *Three examples of an information window for the selection in Figure 4.4. Top left: the* MANET *software, with additional group information. Top right:* Mondrian. *Bottom: a report window in* Data Desk.

conditional on the values of all the others. In the example, this would require two: survival by gender given class and survival by class given gender. Doubledecker plots (Hofmann and Wilhelm; 2001) have been used in Figure 4.5. Instead of alternately splitting the x and y axes as in a standard mosaic plot, only the x axis is used. This allows the direct comparison of all highlighted proportions. The upper plot shows that male survival rates were lower than female survival rates in each class. The lower plot shows that female survival rates declined by class whereas male rates did not — second class males had a lower survival rate than those in the third class. Notice how much extra information the complementary plots provide. It is difficult to judge the survival rate of females in first class in isolation.

Keeping track of a selection from a single plot is easy, but it is far more difficult to manage selections from multiple plots. In principle, an information window could be automatically created to report and manage the current selection, but it is not easy to see how this could be done in a readily interpretable way. An SQL-equivalent would be technically correct but only comprehensible to experts. A computer-generated text is likely to be clumsy and awkward to interpret. Incidentally, but importantly, one task that the software can easily perform, which is extremely helpful, is to report the absolute and relative case counts for selections (see Figure 4.6). Selections across several variables can sometimes shrink to nothing, even in huge datasets.

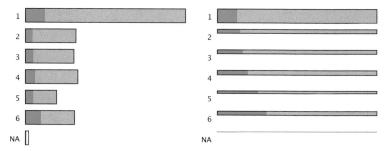

Fig. 4.7. *The Bowling Alone gun question displayed as a barchart and spineplot (with the same selection as in Fig 4.4: those who agree generally/definitely they would do better than average in a fistfight).*

4.3.4 Varying Plot Characteristics

There are rarely single "optimal" displays in data analysis. Information is gained from working through a range of potential displays: rescaling, resizing, zooming, reordering, reshading — reformatting them in many different ways. Point sizes may be made bigger in scatterplots to draw attention to outliers. The aspect ratio of a plot may be varied to emphasize or to downplay a slope. The choice of aspect ratio affects how mosaic plots look, and different ratios may be appropriate for different comparisons within a plot. Barcharts may be switched to spineplots to compare highlighted proportions better. Figure 4.7 shows how the barchart emphasizes absolute numbers and the spineplot emphasizes rates. Boxplots may be switched to dotplots to check for gaps in the data. A scatterplot axis may be rescaled to exclude outliers and expand the scale for the bulk of the data. The variables in a mosaic plot may be reordered to change the conditional structure displayed. α-blending may be applied to a parallel coordinates display to downplay the non-selected cases (Figure 4.8). The binwidth of a histogram may be varied to provide a more detailed view of the data (Figure 4.9). A set of plots may be put on a common scale to enable better comparisons.

Data may be sorted. There are surprisingly many ways that this can be done. A chart of US States from the *Bowling Alone* dataset might be sorted by the number of high income respondents, by the proportion over 70, by any one of a number of criteria. Figure 4.10 shows two examples, though clearly in this printed form they are unsatisfactory — there are too many States for proper labeling. A horizontal layout might be used if a presentation graphic is needed. Presented interactively, there is, of course, no problem, since users can then query, zoom in and, if they want to, resort.

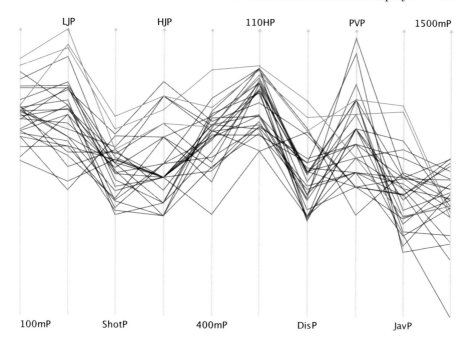

Fig. 4.8. *Parallel coordinate plot of the decathlon points at the Athens Olympiad 2004 with the medal winners highlighted.*

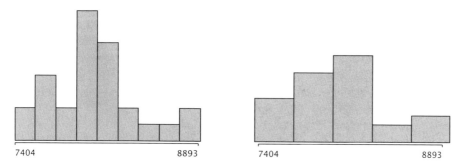

Fig. 4.9. *Two histograms of the numbers of points the competitors achieved in the Decathlon at Athens. The second plot was obtained by varying the binwidth using a direct manipulation keyboard control.*

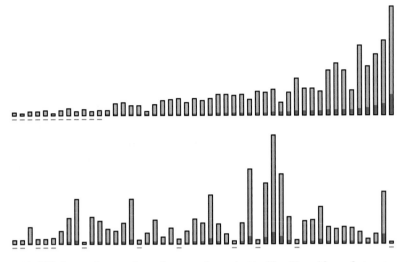

Fig. 4.10. *US States by number of respondents in the Bowling Alone dataset sorted above by the number of higher income respondents (California is first and the noticeably smaller state in 6th position is New Jersey) and below by the proportion of respondents over 70 (DC has the highest value, then Florida and Nebraska).*

4.3.5 Interfaces and Interaction

Successful interactive applications need good interfaces. Whether a software's interface is really excellent or not is likely to be a subjective matter. Most of us can tell when an interface is poor, though as is common with design issues that does not mean that we can specify how an interface should be. There are many different possibilities, and both individual options and the overall consistency of an interface are relevant. Different people work in different ways and it is useful to offer a range of possibilities. There are good discussions of the issues in Nielsen (1994).

For interactive graphics it is important that the options to work with the displays are readily available (not just hidden away in menus) and that they do not clutter up the screen, taking valuable space away from the displays themselves. Frequently used options should be easier to find than ones that are rarely used. Ideally, users should know intuitively what they can do with any given statistical object, whether the commands are visible or not. This is unreasonable to expect for all but expert users, so both help facilities and command redundancy (providing access to commands in more than one way, e.g., both as shortcuts and in menus) are necessary.

The more flexibly and more easily tasks can be carried out, the more likely they will be, and this implies the need for interactive controls. Some may be made implicitly available in a cue, so that they appear when the cursor moves across their location. For instance, it is easy to switch between barcharts and spineplots using a command located along the base of the display. Some may be made discreetly available like the anchorpoint and binwidth slider controls for histograms in MANET.

The best way to make commands available to users is a delicate interface question. It depends on how frequently the command may be needed, on the user's level of knowledge of what might be worth doing in any situation, on what kind of help is provided, and on how often the package is used. There will clearly be no one best way, and the most sensible approach is to provide access in more than one way: shortcuts for the experts, pop-ups to offer guidance, and menu items as backup giving an overview of all commands.

The object-verb paradigm (select the object you want to act on and then what you want to do with it) is the natural approach for exploratory work. It also makes it easier to offer the user support with screen hints or pop-up menus, as for most objects there are only a limited number of possible options. Commands may be provided in a variety of ways, ranging in decreasing immediacy of interaction from *cues* over *keyboard shortcuts, sliders, pop-up menus, floating controls (palettes or dialog boxes), pull-down menus, modal dialog boxes* to *command lines*. The amount of information that can be specified with these controls ranks in reverse order: cues just allow a binary switch, keyboard shortcuts a single action, sliders control a range of values, palettes and pop-ups offer command selections, while menus and dialog boxes offer the full range of user interactions. Floating controls remain open while the changes entered in them take place. With a modal dialog box, the dialog takes control of the process and changes are only made after the box has been dismissed with "OK" or a similar command.

Exploratory data analysis (EDA) has to be flexible and fast. There are many ideas to be checked, and analysis should neither be held up nor distracted by the interface. This has two implications: firstly that frequently

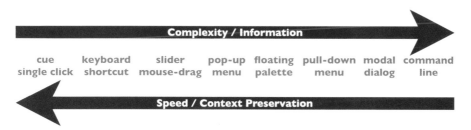

Fig. 4.11. *Hierarchy of user interface controls according to their information content, complexity, speed, and context preservation.*

used commands should be on hand when required and secondly that users should be warned if displays may mislead.

Cues are one way to make commands accessible. They are areas within a display where the cursor changes form to indicate that clicking in the area will change the display. The disadvantages of cues are that users may not know they are there and may not know what they mean, if they do find them. The advantages are that they are quick and unobtrusive. Examples in MANET include switching between barcharts and spineplots, switching axes in scatterplots and rotating dotplots through 90° (i.e., switching from vertical to horizontal displays or vice versa).

4.3.6 Degrees of Linking

Changes in one display often have impact on other displays. Velleman incorporated a useful classification of linking between displays in Data Desk many years ago:

Hotlinking means that any change in one window is propagated immediately to other windows hotlinked to it.

Warmlinking means that the user has the option of propagating the change for each linked window separately.

Coldlinking means that after the initial link between the windows, no further connection exists between them.

The main example of hotlinking is selection. There can be few situations, if any, when highlighting the selected points in other plots would not be desirable. Linking display changes is a more subtle issue and warmlinking comes into play. Commonly, it is sensible to link category order in barcharts and mosaic plots, especially as that is the most elegant way of varying a variable's category ordering within a mosaic plot. However, it may be informative to compare barcharts of the same data with quite different orderings (cf. Chapter 5).

Coldlinking is used when graphics have been prepared for publication and are not to be changed again. It is an option in spreadsheets when values instead of functions are copied.

Warmlinking is useful for scaling of continuous variables. If a variable's scale is changed in one window, then it can be useful to change it in all. The concept of warmlinking arises in an important way in another context. Nowadays many datasets, especially large ones, are continually developing. The term data streaming is used, where more and more data are collected on a continuous basis. Hotlinked graphics should then also change all the time, and this could be mildly distracting and potentially misleading. Warmlinking to the dataset is the best solution. People who work with databases are more aware of these issues than statisticians, most of whom have up till now been accustomed to work with fixed datasets.

4.3.7 Warnings and Redmarking

Displays may mislead if scales are chosen such that some points are not included or if the screen resolution is not high enough to display areas of small numbers of cases. Redmarking (see Hofmann (2000) and Section 3.3.1) is one way to counteract this. A red mark is drawn on the display, ideally positioned to indicate where the problem arises. For histograms and barcharts, a line can be drawn underneath the affected bar (because the bar as a whole has too few cases to be visible or because the number of highlighted cases is too small or because the number of non-highlighted cases is too small). For displays of continuous variables (e.g., histograms and scatterplots) with scales that do not cover the range of all cases, some sort of warning hint must be shown, be it a red square drawn in the corner of the display or a red frame, indicating that some points are plotted outside the window.

In some graphics, only the fact that a histogram's horizontal axis appears to extend far beyond the data alerts the analyst to the presence of a few outlying cases. Because it is common in large datasets to have isolated gross data errors, it is important to be able to recognize and deal with them (cf. Figure 3.2).

A warning of some kind is also in order if highlighting proportions may mislead. In Figure 4.10, States where the numbers highlighted are

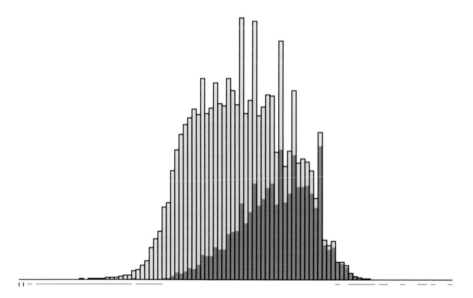

Fig. 4.12. *A histogram with red bars indicating bars that are too small to be displayed at this screen resolution. The data are logged amounts of 450,000 bank deals.*

too small for the highlighting to show have been redmarked. Redmarking would also be used if the non-highlighted proportion was too small to be displayed.

Figure 4.12 shows the data in Figure 3.2 again with one deal type selected. Most of these deals were for larger amounts, but the redmarks under the bars to the left of the middle show that there were a few smaller deals of this type, too. The red marks to the right of the middle show that not all deals for those larger amounts were of that type.

4.4 Interaction and Large Datasets

Some tasks need a refinement of method for larger datasets. Finding an individual highlighted case in a small dataset is just a matter of a quick scroll through the data. This is not a practical option for large datasets, but implementing a command to jump to the highlighted case is a straightforward solution. For multiple highlighted cases, the initial solution would be to show only those selected and to mask the rest. The disadvantage of this approach is that cases adjacent to the highlighted ones are not visible, and it is useful to see the selection in context to interpret it properly. Checking missing values in the Tropical Atmosphere Ocean dataset (http://www.pmel.noaa.gov/tao/), there were some unusual values before the missings, suggesting a developing problem. Context is a crucial word for large datasets. For small datasets, the whole dataset is the context and it can be viewed all at once. For large datasets, this is simply not possible.

Above all, the distinction between point-based displays like scatterplots and area-based displays like histograms or mosaic plots is important, because area displays scale up in an obvious way, whereas point displays do not (cf. Chapter 3).

4.4.1 Querying

Querying or selecting a bar in a histogram (or a cell in any area display) looks much the same whether the bar contains 10 cases or 100,000 cases, but the scale of information is different. Knowing that 10 cases are aged 24 and that 6 are female (because they are currently highlighted from a selection of all females) and the rest are male is about the highest level of detail the data can support. With 100,000 cases, much more can be done:

- Is the proportion of females significantly different from 50% (or from the proportion in the whole dataset)?
- Does the proportion of females rise or fall monotonically across the range?
- What does linking to additional variables reveal?

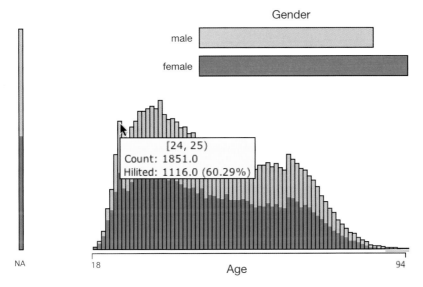

Fig. 4.13. *The age distribution of Bowling Alone respondents with females high-lighted.*

Figure 4.13 shows the age distribution from the *Bowling Alone* dataset with females highlighted. The query shows that there are 1,116 females (60.3%) out of 1,851 respondents aged 24.

If data are not visible, they cannot be directly queried. If redmarking is used (cf. Section 4.3.7) then a consistent solution is to interrogate the warning red marks themselves.

Querying point-displays from large datasets is more difficult than for small datasets, because the overwhelming majority of points will not be visible as individual objects. Outliers may still be queried, but there is not a lot to be gained by querying the middle of a solid mass of points and discovering that the screen pixel picked out is overlapped by 14,758 points of which 3,821 are currently selected. In practice, scatterplots of the full dataset will likely be displayed using either some sophisticated form of density estimation or just simple binning (cf. Section 3.5). Standard querying will only be of value for outliers or after zooming into a subset of the plot.

4.4.2 Selection, Linking, and Highlighting

In large datasets, you will want to do more than making simple selections in a single plot. Having selected those with high income, you may want to explore differences by sex, by age, and by other covariates. This means that the selection process becomes more complicated and it becomes increasingly important to make sure that appropriate comparisons are made.

For large datasets, many selections may be interesting concurrently. Starting with one selection (say gender), several paths could be followed up in parallel to discover which seems most promising. In the income example, it is not immediately clear which of several explanatory variables might be most important, and they would have to be explored in several different ways.

These are more complex procedures than are necessary for exploring small datasets, where it is easy for an analyst to maintain an overview of the whole process. Software is needed that provides support for data exploration. There is a danger of analysis being handicapped by having to perform housekeeping tasks to avoid the risk of losing track of analyses. A more structured approach is necessary for large datasets than for small ones.

There are two additional problems associated with linking and large datasets — visualizing small numbers and highlighting in scatterplots. With small datasets, it is unusual not to see every case in every graphic. With large datasets, small groups of cases will be invisible due to insufficient screen resolution. Redmarking (see Section 4.3.7) alerts us to the problem but does not provide a full solution. One alternative could be a local logical zooming and that would work well as long as the overall context is either not necessary or is maintained in a bird's eye view. Further alternatives are censored zooming and quantum zooming discussed in the next section. Hotselection, suggested by Velleman and implemented in DataDesk, is an attractive option, especially for large datasets. Only currently highlighted points are shown (Figure 4.14), so it is like a drilling down, except that changing the selection immediately changes all hotselected plots and analyses, so that you can switch between different groups

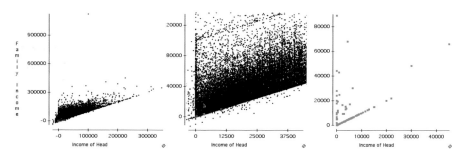

Fig. 4.14. *With only a few cases highlighted (all younger than 18), any structure of the highlighted cases is hard to see (left). Simple zooming can help (middle), and hot selection (right) helps to focus on only the selected subgroup.*

of cases very easily. For comparing groups in this way it is advantageous to freeze the scales in the plots so that they remain properly comparable.

Scatterplots were the first display for which linking was shown (McDonald; 1988) and in a way it is surprising that the inherent problems have never been resolved. Displaying multiple points at the same location is still not handled by many software packages, even though it was a feature of the original 3-D rotating plot system PRIM-9 (Fisherkeller et al.; 1971). What to do about highlighting in glyph-based plots is a much more difficult problem. In Figure 4.15 (left), it looks as though most data are highlighted, but in fact less than 4% are. This is because the highlighted cases are plotted on top of the non-highlighted cases. This is a sensible ordering for smaller datasets but leads to the unsatisfying results in Figure 4.15 (left) — a problem that must be addressed in all glyph-based plots for large data; see also Chapter 6. Figure 4.15 (right) uses α-blending, as described in Section 3.5, for both the data and the highlighting. Now it is easy to see that only a small fraction of the cases are selected, but the resulting image is still hard to interpret quantitatively.

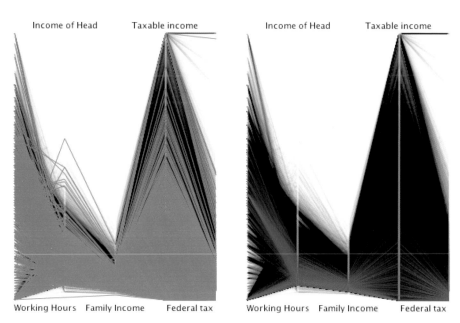

Fig. 4.15. *Parallel coordinate plots without α-blending on the highlighted cases (left) and with α-blending.*

4.4.3 Varying Plot Characteristics for Large Datasets

Zooming

As Shneiderman's mantra puts it: "Overview first, zoom and filter, then details on demand" (Shneiderman; 1996). Zooming is an important tool of graphical analysis. It may be a simple idea in theory, but it is one that turns out to be complex in practice. Zooming does not only consist of image magnification. The larger the dataset, the more zooming is needed. For large datasets, this is not just a matter of a zoom factor of 2 or even of 10, it can be a factor of several powers of 10 and it may not be symmetric in the axes. In Figure 4.16, there are two plots of profit against amount for some 450,000 company bank deals. The first shows all the data and the second the plot restricted by a factor of 10^{-5} on the x-axis and 10^{-3} on the y–axis.

To decide just which levels of zoom reveal information in data requires more sophisticated tools than are available at the moment. Zooming should be guided by user-specified data-driven criteria with multiple levels of zoom and multiple zooms in parallel. You have to be able to move between different levels smoothly and without losing a sense of context, and you need to be able to compare parallel zooms (i.e., zooms of the same level at different locations in the data); for instance, is the pattern of income in the late 1970s the same as that in the 1980s and the early 1990s? Orientation and navigation tools are needed to keep track and to guide. Current solutions that use a bird's eye overview in a separate window or non-linear transformations such as fisheye are effective, especially when they include interaction, such as being able to pan around in the bird's

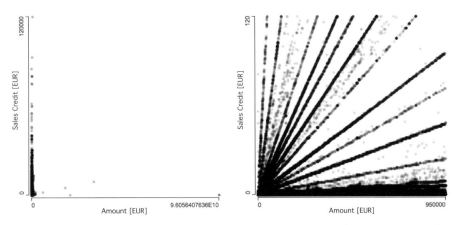

Fig. 4.16. *Scatterplots of profit against amount. On the left, all 450,651 cases. On the right, the 261,636 cases close to (0,0).*

Table 4.1. *Methods of Zooming*

Method	Description
Basic	The display is magnified.
Logical	As magnification increases, more detail is shown.
Object	Only certain object types are magnified.
Semantic	Magnification depends on information to be imparted.
Ceiling-Censored	Objects are magnified up to a limit C.
Floor-Censored	Objects smaller than size F are not displayed.
Quantum	Highlighted small groups are magnified.

eye view and being able to zoom in and out continuously and smoothly. A parallel slider control for the level of zoom can be incorporated in the overview display. For high levels of zoom, it is possible for the exact zoom location to be too small to be shown at the resolution of the overview window, but a (red) marker identifies the location.

Different kinds of zoom are necessary too. Zooming is not just magnification for large datasets, it entails showing different levels of detail — the term logical zooming (where there is an obvious default for what additional detail should be shown on zooming in) or semantic zooming (where the level of information shown as you zoom in depends on the meaning to be imparted) may be used in this case. (See also Stolte et al. (2002) for more on these issues.) A further alternative, object zooming, is to zoom only certain kinds of object, for instance either just the edges in a graph or just the vertices.

Censored Zooming

Cells with small counts in mosaic plots need to be dealt with in other ways. One approach, censored zooming, has been introduced by Hofmann for fluctuation diagrams, a version of mosaic plots (see also Section 5.4.5). Each cell in a fluctuation diagram is allocated the same amount of space, and the cell with the maximum frequency fills its space completely, thus fixing the scale for the rest of the diagram. You can then interactively adjust the scaling so that all cells with frequencies higher than a ceiling value fill their spaces completely and the remainder are drawn in proportion to their frequencies. By progressively lowering the limiting ceiling value, more and more of the smaller cells become visible. For ceiling-censored zooming, the affected cells are bordered in red.

Consider a fluctuation diagram of four variables (Marital Status, Region, No. of Children, and Gender) of the US Census dataset with 5, 4, 10, 2 categories, respectively, making 400 cells in all (Figure 4.17), of which 146 are empty. For large datasets, the cell-occupancy distribution is often

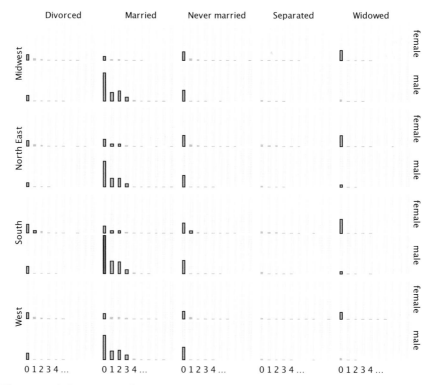

Fig. 4.17. *A fluctuation diagram from the Census dataset. Marital Status × Region × No. of Children × Gender. Many combinations are empty or rarely occur.*

highly concentrated with most of the data in some 10% of the cells (in this case, more than 80% of the data are in fewer than 10% of the cells — 39 combinations). Assuming that a window of about 200 by 300 pixels were used, then each cell could have a maximum of 70 pixels. Any cell with a frequency of less than 1/140th of the maximum (assuming rounding to the nearest pixel is possible) would not be drawn. Because the difference between empty cells and non-empty cells can be an important distinction, it seems advisable to give non-empty cells at least a pixel, but this means that "small" cells cannot be distinguished from one another.

The biggest cell in Figure 4.17 has around 7.3% of the data, about 4,600 cases, so that no cell with frequency less than 30 would be drawn with more than a pixel. Slightly bigger cells would be drawn with similar size. Lowering the ceiling value to 235 (i.e., that any cell with 235 or more cases completely fills its 70 pixels) ensures that the smaller cells can be compared visually with one another, see Figure 4.18.

Invisibility of small cells is not necessarily a disadvantage, as they may otherwise distract from major data features, and another form of

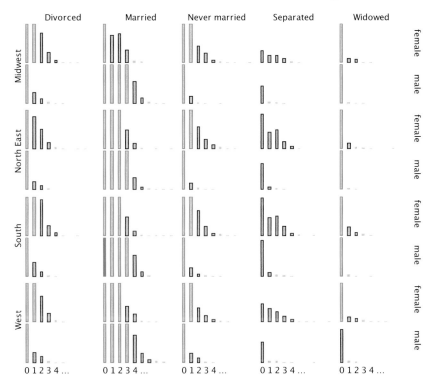

Fig. 4.18. *Ceiling-censored zooming. The same plot as Figure 4.17, but with a ceiling of 235. Cells with more than 235 cases are bordered in red. Now the distributions for "Separated" are clearly visible. Before they could not be distinguished.*

censored zooming, floor zooming, is used to mask all cells below a specified "floor" size.

Finally, it may be informative to compare small levels of highlighting and this would benefit from yet another form of zooming. Suppose that rates of marital separation in different subgroups of the population are being studied. The difference between 0.5% and 1.5% is relatively big, but yet barely visible, if at all, see Figure 4.19. These differences have to be displayed so that they can be judged without losing context. Quantum zooming would be an appropriate name (now all that is needed is to design and implement the method).

Shneiderman's mantra, quoted at the beginning of this section, is one central part of the story, but zooming has a bigger role to play than that. The larger the dataset, the more zooming is likely to be needed, and the more different kinds of zooming are needed.

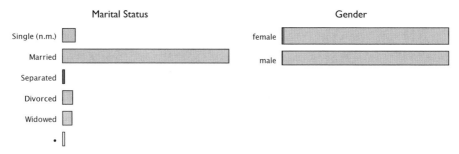

Fig. 4.19. *The 855 (ca. 1%) of respondents in the Bowling Alone dataset who are recorded as being separated have been selected in the plot of marital status on the left. The rate for women (1.5%) is three times that for men (0.5%) (right-hand plot).*

Sorting

Although sorting is a basic idea, it is an extremely powerful tool. You can sort the categories in a bar chart, the axes in a parallel coordinates display, or the variables in a mosaic plot. But how should anyone sort a display? Detailed provision using command lines or dialog boxes is all very well (and command lines are essential for repetitive tasks), but what about sorting "on the fly"? One way to sort categories is by the current selection according to the graphic form, either barchart to sort by absolute values or spineplot to sort by proportions. The reasoning is that the sorting you want to see is the one associated with the current selection. The follow-up is that you need to know on what criteria a particular display has been sorted and this information should be automatically recorded and included.

Sorting options are taken for granted in many statistical operations, and for small datasets you can — any sorting algorithm will sort a small number of cases instantaneously. When datasets become large, sorting methods become more of an issue. Any old algorithm is not enough and speed is of the essence. Whether it is sorting continuous data for highlighting in boxplots or reordering categories in barcharts or mosaic plots, the results should be available instantaneously and comprehensibly.

Large datasets affect sorting in other, more critical ways. Any display for a variable with many categories will have to be sorted. Boxplots drawn side-by-side for three different groups can be easily compared by eye, whatever the ordering of the groups. When there are even half a dozen groups, this becomes much more difficult and with 20 groups impossible. Figure 4.20 shows some logged financial transaction data in boxplots by branch for just under half a million transactions. It would be helpful to sort them by median or possibly by maximum and to take account of the number of cases in each group (which here varies from 28 to 190,000!).

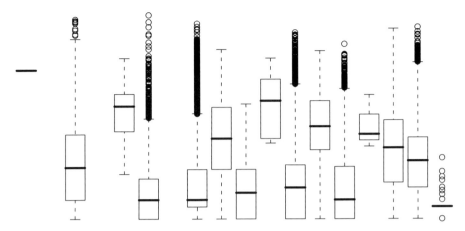

Fig. 4.20. *Logged amounts of financial transaction data in boxplots by branch (for reasons of confidentiality, the branches are not labelled)*

As in Figure 4.20, there are more sorting possibilities with large datasets. Figure 4.10 showed two sorting options for a barchart with almost 50 categories (Alaska and Hawaii were not surveyed, while DC was). Even more sortings can be considered if the possible weightings of the barchart are taken into account. The data could be weighted by income, by age, by number of children, or by any other available (appropriate) weight. Given so many alternatives, sorting needs to be carefully managed and monitored so that the user is always informed of current orderings.

Having a large number of variables leads to new sorting challenges. The list of variables itself needs to be sorted (probably most profitably by topic, but also possibly by statistics such as the proportion of missings). Displays of multiple variables such as mosaic plots and parallel coordinate plots have to be sortable. Any collection of displays on-screen will be more informative if they can be arranged in a structured way.

Common Scaling

Another way to enhance a collection of displays of variables is to use common scaling, i.e., give comparable axes the same scale. This can mean ensuring that the same scale is used for all displays involving a particular variable or choosing to display a set of different variables of the same type with a common scale. This is a natural option for parallel coordinate plots, where variables are often of identical types. A prime example can be found in Unwin et al. (2003), where residuals from different models are plotted together on over 300 axes.

Multiple Views

Insights are gained by looking at data from several different points of view. Features that may be apparent in one view may not be visible in another. If all the information in a large dataset is to be displayed, lots of different views are needed. Eick and Karr (2002) have written of perspectives, by which they mean sets of views of the data that can be designed for specific purposes. This is a valuable approach for repetitive tasks (such as daily reviews of investments or regular sales reviews), but here the task is exploring data, interactively working with multiple views: choosing, organising, scaling, and, of course, linking.

The Fistfight question from the *Bowling Alone* dataset is displayed in Figure 4.4. There are many different variables that might be related: demographics like gender, age and marital status; sociological ones as recorded by questions on social habits; but also attitudinal ones as expressed in opinions on other issues. To be able to filter through these quickly and informatively is all to do with organisation and management: being able to resize, reorder, realign many graphics simultaneously; being able to group related graphics together; being able to combine groups of variables together. Combining variables through transformations and recodings can be valuable, especially when the results can be swiftly and immediately assessed with graphics. It is then readily apparent whether a new variable is going to convey additional insights and what adjustments might be worth trying to increase its contribution still further.

4.5 New Interactive Tasks

Much more could be done interactively in data analysis software than is done at the moment. Many tasks that are particularly relevant for large datasets could be made faster and more flexible if made interactive.

4.5.1 Subsetting

It is not always necessary to look at the whole of a large dataset. A sample may be enough for checking overall structure. Some subsets may deserve deeper individual analyses quite separate from the rest of the dataset. Defining subsets is a naturally interactive action and implies keeping track of them, ensuring that appropriate comparisons are made for them, and managing the analyses. The efficient organisation and management of variables and analyses is not an interactive task in itself, but only if this is effectively achieved does interaction become feasible.

4.5.2 Aggregation and Recoding

Aggregation and the construction of new variables based on the results of analyses are other frequent tasks in the analysis of large datasets that benefit from interactive support. Grouping categories of a variable together is necessary when rarely occurring categories are to be grouped together into "others" or when the dataset contains multiple options for the same response. Writing recoding statements is a labour-intensive way of tackling this, and interactive methods allowing the user to drag groups together are both simpler and more direct. Studying survey responses by US State in the *Bowling Alone* dataset might suggest aggregating the smaller States into a single group or combining them with geographic neighbors. Sorting and grouping from a barchart or selecting using a map are both obvious ways of doing this but imply that the software should provide an easy and intuitive option to store the result as a new variable.

4.5.3 Transformations

Transforming variables is a common task in all data analyses, but when few variables are involved it is a simple task with at most some mildly irritating recoding to do (for instance, combining the four race variables in the *Bowling Alone* dataset into one). When there are many variables and many potentially interesting transformations (e.g., discretising the income variables in different ways, creating summary scores on related social variables, combining income and housing variables to produce a wealth measure), then the burden is far greater. The effort lies in searching for the most useful combination of variables and the most informative transformations, not in specifying one particular formula. This should be an interactive task, but current software implementations lag well behind what would really help, probably because of the level of complexity involved.

4.5.4 Weighting

Large survey datasets are often based on some form of unequal sampling so that the cases are weighted (e.g., in the PISA study, Adams and Wu (2002)). Any statistics calculated have to take account of the weights and so should any graphics drawn. With area plots, such as histograms and barcharts, the heights of the bins or bars should be drawn proportional to the sum of the weights of the cases in the class. For point plots, such as dotplots or scatterplots, the area of the point representing a case can be drawn proportional to the weight. This leads to overlapping and overplotting, which can be partially dealt with using alpha-blending or a form of density estimation (cf. Section 3.5).

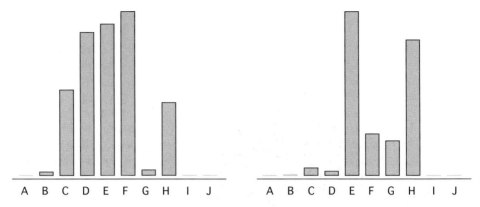

Fig. 4.21. *Barcharts of classes of bank deals. On the left, the numbers in the classes. On the right, the classes are weighted by turnover.*

Large datasets may be weighted for other reasons, too, and the same data may be informatively weighted in different ways: by amount, by number of items, by complexity, by whatever nonnegative, additive feature you wish. Figure 4.21 shows two barcharts from the company bank deals dataset. The left plot displays the numbers of different classes of deal. The right plot displays the same deal classes weighted by turnover. Unfortunately, for reasons of confidentiality, the actual class names cannot be shown, but the differences between the relative heights in the two diagrams are striking. Note that the vertical scales are not directly comparable here. This will always be the case when comparing two diagrams with different weightings.

Further examples of these kinds of weighted plots can be found in Chapter 9 (Figure 9.9 and Figure 9.11) and Chapter 11 (Figure 11.5 and Figures 11.11 to 11.15).

With data recorded by spatial region, it is useful to be able to display variables like party election support weighted by the corresponding electorate or illness rates weighted by population. Occasionally, datasets may be summarised as data cubes, and the cell counts are the weights for their respective cells. However weights arise, it is necessary for graphics to be adjusted accordingly, and switching between weightings ought to be an interactive process, so that the effects are clearly seen. MANET uses drag and drop for this purpose, which is a natural way of going about it.

4.5.5 Managing Screen Layout

Larger datasets require more displays, as discussed in Section 1.5.6. Exploring many variables demands many univariate displays and, needless to say, many more multivariate options. Scatterplots, trellis plots, mosaic plots, and parallel coordinate plots take up a lot of screen space, and so do micromaps and all "small multiples" displays. Several packages tile windows to provide an overview, though they then treat all windows equally, so that a barchart of two categories and a scatterplot of thousands of points get allocated the same amount of space. Managing and organising sets of displays efficiently, varying the layout for comparisons, and resizing individual plots to emphasise interesting features all benefit from software support.

None of these tasks is inherently graphical, but all would be easier to tackle with interactive user interfaces and with graphical outputs to monitor the results. It is necessary to be able to switch the focus of analyses smoothly and intelligently while maintaining both the context of the detailed analyses and an overview of the whole task.

4.6 Summary and Future Directions

To a great extent, the recommendations in this chapter amount to asking for more of everything for dealing with large datasets:

- more powerful querying;
- more forms of zooming;
- more complex selections;
- the ability to carry out multiple investigations in parallel (multiple selections, multiple subsets, multiple zooms).

None of these capabilities are primarily statistical, but they provide essential support for effective data analyses of large datasets.

Some interactive tools do not scale up as well as others, but this does not present major challenges. Most of the data analysis that can be done with interactive graphics for small datasets can be carried out for large datasets, too. The real challenge lies in adding interactive capabilities to tackle the new organisational and management tasks that arise with large datasets.

Part II

Applications

5

Multivariate Categorical Data — Mosaic Plots

Heike Hofmann

Seid umschlungen, Millionen![1]

Friedrich Schiller, *An die Freude*

5.1 Introduction

Categorical displays are fortunately relatively independent of the number of cases they display. Unfortunately, the number of categories of a variable can grow with an increase in the number of cases. The challenge for categorical displays is to deal with large numbers of categories and with variables with skew marginal distributions. The goal in mind must be — as always with visual displays — the representation of both global trends and local features, individuals or small groups not fitting any general pattern. In an interactive framework, techniques like reordering and grouping of categories help to produce clearer images of "what is in the data". Querying mechanisms and display variations are used for extracting summarized information.

5.2 Area-based Displays

Categorical data are best displayed in area-based plots. Each cell or bin of the data is represented by a rectangular area, the size of which is — in some way — related to the number of cases in the cell it represents. Mostly this relationship is linear but there are useful variations.

Area-based plots are fairly stable with regard to the absolute numbers of cases they represent. As long as the relative proportions between cells are not affected, the diagram will look exactly the same regardless of the total number of cases. The difference is that individual cases are

---

[1] Be embraced, you millions!

represented by smaller areas for higher numbers of cases. This may become a problem in a data analysis when the focus is on an individual or
on a small group of individuals. It can happen that the height or width
of an area representing a small group would be represented by less than
1 pixel of the display, so that the group is not visible unless redmarking
(cf. Section 4.3.7) or some other action is taken. The idea is to protect the
analyst from jumping to false conclusions.

Once this problem is identified, there are several solutions possible.
Figure 5.1 shows barcharts of 700 cases on the left and 700,000 cases on
the right. The proportions between categories are equivalent, yielding the
same overall picture, i.e., category a is about twice the size of category b,
and half the size of category c. In both barcharts the highlighting shows
all but 20 cases in bin a (corresponding to 10% of the cases in the left
barchart side but only 0.01% in the right), 5 cases in bin b, and 1/4 of
the cases in bin c. Because not all cases in bin a in the right barchart
are highlighted even though it appears to be that way, a warning line
has been added underneath the bin. Bin b on the right is marked the
same way. Here, some cases are highlighted but the area representing
this group is too small to be drawn.

Another solution to the problem of screen resolution would be to fix
a minimal height of 1 pixel whenever at least one case is highlighted (or
only one is not highlighted), thereby graphically over-representing this
piece of information in order not to neglect it. This is only practical, when
the bars themselves are big enough.

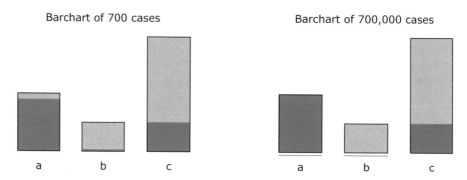

Fig. 5.1. *Barcharts of 700 cases (left) versus 700,000 cases (right). Proportions
between categories are the same, yielding the same overall picture. Highlighted
are all but 20 cases in bin* a, *5 cases in bin* b, *and 1/4 of the cases in bin* c. *The 20
cases in the right barchart represent less than 0.03 %oo and correspond to an area
with a height of less than one pixel — this makes a warning line as shown below
categories* a *and* b *necessary.*

5.2.1 Weighted Displays and Weights in Datasets

A further issue when working with categorical variables in large datasets is the format in which the data are stored. It can be efficient to just store case counts for all the possible combinations (like in a data cube) instead of storing each case individually, thereby saving both storage space and main memory without any loss of information. The (in)famous Titanic data (Dawson; 1995), for example, is reduced from 2201 individual cases to only 24 non-empty combinations of the four variables. This format is commonly used in statistical modelling — it is particularly useful if the sample size is large and all variables involved are categorical. Sparse tables also benefit from using a weighted data format. With increasing numbers of variables and categories, it loses its advantage over a mere listing of individual cases. It is the trade-off between sample size N and the overall number of categories $c_1 \cdot c_2 \cdot \ldots \cdot c_p$ (where c_i is the number of categories of variable X_i) that has to be considered. This number of categories is further reduced in a weighted data format by one for each combination of categories that have a zero count.   Table 5.1 shows the trade-off between sample size and number of categories for three very different datasets.

Software should be able to deal with the data either way.

5.3 Displays and Techniques in One Dimension

With increasing numbers of cases, the numbers of categories tend to increase, too. This can even happen for variables like *gender of policy holder* (see Figure 5.2), which you would think could only have two categories. "Company" and "not known" are understandable; "other" was because the dataset had not been cleaned first.

With increasing numbers of cases, the possibility of individuals not fitting a predefined categorization grows. Usually, the number of those individuals will be small compared to the overall number of cases. With nominal variables the number of categories increases quite naturally with

Table 5.1. *Sample Size and Number of Categories for Three Different Datasets. (See (Whittaker; 1990) for more information on the Rochdale data.)*

Dataset	Sample size	# categories $c_1 c_2 \ldots c_p$	# empty combinations	Ratio of reduction
Titanic	2201	32	8	2201/24 = 91.7
Rochdale	665	256	165	665/91 = 7.3
Internet Usage	10108	5.55×10^{36}	?	1.0

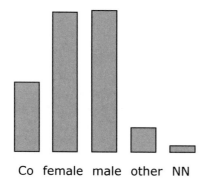

Co female male other NN

Fig. 5.2. *Barchart of policy holder's gender in a car insurance dataset. With an increasing number of cases, the number of categories increases. Besides the obvious two categories, there are the categories* company, other, *and* not known.

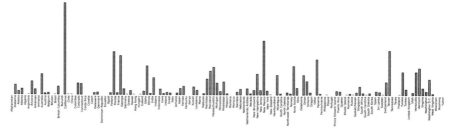

Fig. 5.3. *Barchart of* Country, *categories are in lexicographic order.*

the number of cases. Figure 5.3 shows a barchart of the variable *Country* (see Section 1.8.3 for more details on the Internet Usage Data (Kehoe et al.; 1997)). The bins are very different in size — a few large categories dominate the picture. The high variability in category sizes, resulting in a "ragged" looking barchart, is a typical problem of displays of variables with many categories.

It is clear that barcharts like this are of limited use in an analysis — at least in their default form, if the 'usability' of barcharts is based on two properties:

- the ability to get an overview of the marginal distribution,
- the possibility of comparing binsizes, including the possibility of deriving a ranking of the categories from the display.

In the example of Figure 5.3, it is not even easy to tell which of the categories has the third biggest number of cases.

Various methods exist to make barcharts with a large number of categories more readable. The most natural approach is to virtually increase the size of the display by using *scrolling* or the more elaborate *zooming* and *panning* as introduced by Eick and Karr (2002).

Other approaches for making barcharts with large numbers of categories more readable are based on modifications of the *scale of the vari-*

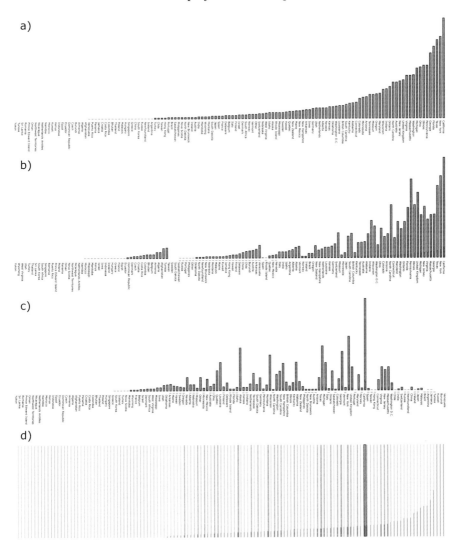

Fig. 5.4. *Four views of the same information: (a) – (c) Barcharts and (d) a spine-plot of* Country *with users stating an overall household income of over $100,000 highlighted. The plotting order of the countries is (a) sorted by number of users per country, (b) sorted by number of users stating an overall household income of over $100,000, (c) and (d) sorted by the percentage of users stating an overall household income of over $100,000.*

able. Two main approaches can be identified: the variable's scale can be changed without changing the overall number of categories, using a re-ordering of the categories. This is, of course, only sensible if the variable is nominal. A second approach is to impose a hierarchy on the categories of the variable, thereby reducing the amount of information displayed. Examples for modifications of this second type are *logical zooming* or *grouping*. Changing the scale of an axis makes it possible to include further information in a display.

5.3.1 Sorting and Reordering

Higher numbers of categories make simple actions, such as a comparison of bin heights, harder. And this is even more so if the bins in question are far apart (see, e.g., Cleveland and McGill (1984) for a discussion of easiness of graphical tasks). With just a few bins, the natural reaction would be to "grab" one bin and drag it to a position, where it is easier to make a comparison — i.e., next to the bin it is to be compared with. Manual reordering gets less and less efficient with increasing numbers of categories. Dynamic reordering and sorting mechanisms for categories are called for. Figure 5.4 shows several aspects of the Internet Usage data after sorting in different ways:

- **Sorting by number highlighted**
  This sorting routine includes as a special case "sorting by case numbers for each category", when all cases are highlighted.
  Plot (a) of Figure 5.4 shows the categories in the order of the number of participating internet users per country. Note that all the US states (and all of the Canadian territories) are treated as individual countries — otherwise the US internet users would dominate the overall picture even more. Most users came from California (1021 users corresponding to 10% overall), followed by New York and Texas. The first non-US entry in the list is the United Kingdom.
  In plot (b), categories are sorted by the number of internet users stating that their overall household income was over $100,000 a year. Places one to three on the list remain unchanged, Georgia and Massachusetts have moved up to positions four and five.
  Note that the sorting is hierarchical: if two bins have the same number of selected cases their original order is retained. This causes the sawtooth pattern in the plot, particularly on the left.

- **Sorting by percentage highlighted**
  In plots (c) and (d) of Figure 5.4, the categories are sorted by the percentage of internet users stating their household income to be over $100,000 a year. On the very right of this ordering are categories with small case numbers, led by Venezuela, Burundi and Kuwait (with 2, 1, and 1 users, respectively).
  Plot (c) shows a barchart reflecting the absolute numbers and plot (d) shows a spineplot (introduced by Hummel (1996), see Theus (2002b)

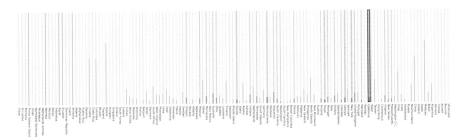

Fig. 5.5. *Spineplot of* Country; *highlighted area shows users stating an overall household income of less than $10.000.*

for details) of the same data. The spineplot reflects this ordering of the bins.

Figure 5.5 shows a spineplot of *Country*. The categories are ordered in the same way as part (c) and (d) of Figure 5.4, that is by the percentage of internet users stating a household income of over $100,000. Highlighted are internet users stating an overall household income of less than $10,000. Among countries on the very right of the barchart — which have a high percentage of users with household incomes of over $100,000 — there are some with high percentages of users stating household incomes of less than $10,000. In this way, a one-dimensional plot visualizes information on more than one dimension. A more general discussion on sorting and orderings can be found in Section 4.3.4.

5.3.2 Grouping, Averaging, and Zooming

Another way of working on the scale of a categorical axis is grouping. Unlike sorting, this method is not limited to nominal variables but works for all categorical — and even real-valued continuous — variables.

Figure 5.6 shows an example of grouping categories in the internet users data. Countries with few users have been combined together into two larger groups, so that the number of categories is reduced and more space is available for individual bins. *Grouping* can also be used to introduce a hierarchical structure to a nominal variable, such as continents for the *Country* variable. Hierarchies can either be given naturally as shown in Figure 5.7, or be imposed by other properties, such as location or size of categories. The multiresolution metaphor (Eick and Karr; 2002) is, for example, based on a hierarchy laid down by the location of cases in the display.

Grouping reduces the number of bins displayed, mostly by reducing the number of smaller categories. When the emphasis is on exploring relationships between categories with small numbers of cases, the strategy has to be modified.

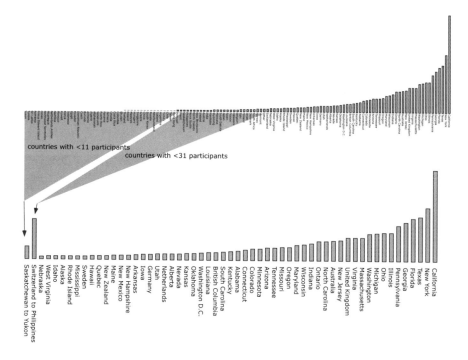

Fig. 5.6. *Barcharts of* Country, *countries with less than 11 participating internet-users are grouped together, the other newly defined group contains all countries with 11–30 participating users.*

Different options exist for focusing on small bins. *Averaging* bins is a variant of grouping bins: when grouping several bins, one new category is produced as a result, combining all the appropriate cases. When *averaging* several bins, one new category is again produced to replace these bins, but instead of all cases only the average number of cases is represented. The graphical dominance of large bins is reduced by averaging, so that smaller bins are emphasized. Technically, this is done by introducing a weighting variable. Averaging is a slightly risky approach since it is non-commutative and non-associative, so that the order in which averaging actions are carried out affects the result. As long as interest is directed only at the non-averaged bins, this is not a problem. Figure 5.8 shows an example from the Internet Usage dataset.

Another option for concentrating on small categories is *zooming* or *logical zooming*. Whereas standard zooming enlarges the displayed graphical elements, logical zooming works on the underlying model and changes it to display more details.

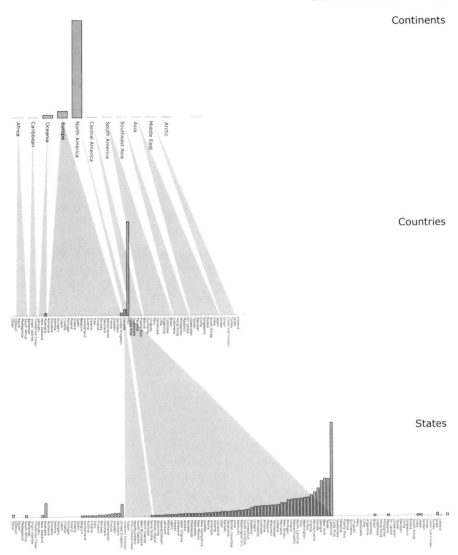

Fig. 5.7. *Repeated grouping of categories can be used to implement a natural hierarchical structure in the data. From top to bottom, the hierarchy is given as* Continents – Countries – States.

5.4 Mosaic Plots

For higher-dimensional displays, the organizational techniques introduced so far become even more important.

When using mosaic plots, there are usually two goals — first of all, trying to find some overall structure between the variables — secondly,

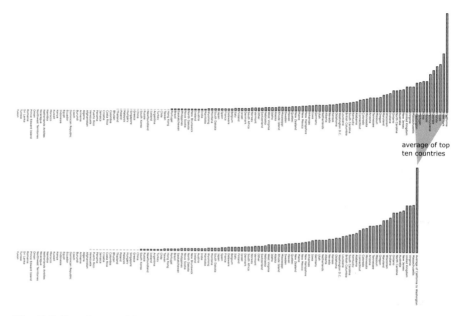

Fig. 5.8. *Barcharts of* Country; *the top ten countries have been summarized in one bin, which displays their average.*

trying to identify combinations of the variables that do not follow the overall structure.

Different variations of mosaic plots (Hartigan and Kleiner; 1981; Friendly; 1994a; Hofmann; 2000) are differently suited for these goals, depending on the specific dataset and the problem. Table 5.4 shows an overview of the most common variations of mosaic diagrams. Typically, for large numbers of categories, grid-based diagrams are easier to read than the default diagram. Additional information can be encoded in mosaic plots by gray-shading the bins or using rescaling methods.

The problems in mosaic plots of variables with high numbers of categories are :

- for higher dimensions, the number of combinations that are empty increases multiplicatively (curse of dimensionality);
- the cell sizes have a skew distribution, making overview diagrams uninformative (a few bins dominate the diagram).

5.4.1 Combinatorics of Mosaic Plots

Mosaic plots are constructed strictly hierarchically — the order of the variables in the plot is crucial for interpretation. When variables in a mosaic plot are reordered, the plot changes in various ways. Even though

Table 5.2. *Overview of the Most Common Variations of Mosaic Plots*

Variation	Binsize	Grid-based	Emphasis on
(a) default	p_{ij}	no	trends, overall patterns
(b) same binsize	1	yes	structure in 'missingness'
(c) fluctuation diagram	p_{ij}	yes	large cell patterns
(d) χ^2 diagram	$p_{i.} \cdot p_{.j}$	yes	deviation from independence

the same cells are shown and their sizes are not changed, the cells' aspect ratios and locations may be different. This happens in a discontinuous fashion, that is, neighboring cells in one setting are not necessarily neighbors in another setting. With p variables there are $p!$ different orderings possible. With 3 variables, this implies a manageable 6 different displays, but with 5 variables, it starts to get out of hand, as there are 120 different mosaic plots possible. For 8 variables, the number is already in the tens of thousands. Obviously, viewing all those displays is both impossible and unnecessary. A good starting ordering is needed and then various additional orderings can be inspected by hand, depending on the dataset.

Figure 5.9 shows an example of how a mosaic plot is affected by reordering its variables. The figure contains three out of the six possible orderings; the same cell is highlighted in all three plots. While the first reordering does not affect the cell much — the main change is a rotation of 90 degrees and a shift to the right, the second reordering also changes the aspect ratio drastically. Even though the size was not affected by the change, it is hard to tell visually. Obviously, interactive methods for reordering variables in mosaic plots are necessary for interpretation. Using highlighting helps to keep track of subgroups of cells during these changes.

Fig. 5.9. *Reordering variables in a mosaic plot affects aspect ratio and location of a cell, but not its size.*

5.4.2 Cases per Pixel and Pixels per Case

Large numbers of variables mean large numbers of possible combinations. A square window of height and width 700 pixels has an area of almost half a million pixels. Drawing a fluctuation diagram with square cells of side a, including the border, and assuming that each cell has a further empty border around it of width 1 pixel (so that there are two pixels between each cell) would allow $490,000/(a+1)^2$ cells. For $a = 9$ you can then display all 4048 possible combinations of 12 binary variables, giving a maximum of 81 pixels for the largest group and allowing the identification of any prominent clusterings in the 12-dimensional space.

As for barcharts and any other area-based diagram, a crucial number for mosaic plots is the number of pixels corresponding to each observation. This number has to be constant for all observations — in the case of a mosaic plot, it is constant for all observations within a bin, but might change slightly between different bins because of rounding. Obviously, the number of pixels per observation is indirectly proportional to the number of total cases displayed. With increasing numbers of cases, the number of pixels per observation is reduced linearly in a mosaic plot of fixed size. Once the number of pixels per observation drops below 1, it is no longer possible to keep track of individual cases. To reduce the risk of drawing incorrect conclusions, additional visual indicators, such as red framing of bins, are needed as a warning.

5.4.3 Calibrating the Eye

To spot something "unusual" in mosaic plots, it is important to know what to expect from each of the variations in a "usual" situation. For a sample of categorical data, the null assumption is commonly Poisson sampling. Figure 5.10 shows mosaics of a contingency table of six variables with three categories each. The cell sizes are realizations of a Poisson distribution with $\lambda = 5000/(3^6) \approx 6.86$. These mosaics are examples of what to expect when there is no structure in the data.

Without structure in the data, the default mode shows a strong grid-like pattern. Nine blocks of cells are visible, each consisting of nine smaller blocks of nine cells each. Perfect independence between the variables yields a perfect grid in a default mosaic.

Variation (d) — the χ^2-diagram — starts from the assumption of total independence between the variables. The cells' areas are proportional to the product of the marginal distributions. To indicate deviation from perfect independence, colour is used: cells with positive residuals (more cases observed than expected) are marked with blue, negative residuals are drawn with red. The height of the coloured area is proportional to the size of the cell's residual. Without any structure in the variables, the colour is distributed across the bins randomly. Patterns in the colour give

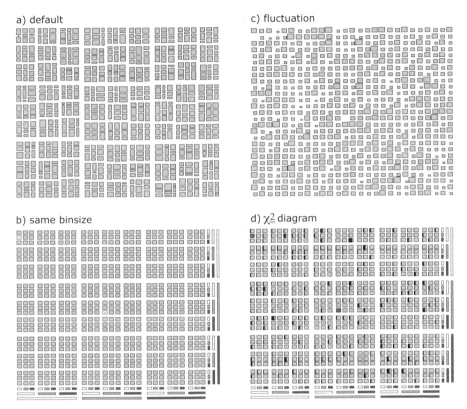

Fig. 5.10. *Sample mosaic plots for each of the variations in Table 5.4. These plots are based on a Poisson simulation, and are therefore meant to demonstrate mosaic plots that do not show any informative structure. The bars to the right and below in plots (b) and (d) label the individual variable values.*

an indication of which interaction between the variables to include in a model.

Variation (b) shows a same binsize mosaic, where all bins have the same size, which emphasizes zero count combinations most. In the example of Figure 5.10, two cells are empty — which is not far from the expected number of empty cells ($5000 \cdot e^{-5000/729} \approx 5.25$).

The fluctuation diagram, variation (c), is closely related to the same binsize variation. Starting from a same binsize mosaic, the rectangles are shrunk according to the number of cases in each combination. That way, the area of each cell is, again, proportional to the number of cases it represents. In Figure 5.10, there is no structure in the cell sizes visible indicating the absence of association between the variables.

Deviations from the expected diagram are now used in the two following examples to find "interesting" cells.

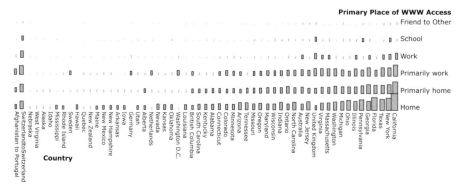

Fig. 5.11. *Fluctuation diagram of* Country *vs* Primary Access. *The categories in each variable are sorted by absolute numbers. Countries with less than 20 users in total have been grouped together.*

Both examples are based on the assumption that there is no dependency between the variables shown. The diagrams make deviations from this assumption easy to spot in two different ways. The fluctuation diagram in Figure 5.11 shows the variables *Primary Place of WWW Access* and *Country*. The categories of both variables are sorted according to the number of participants in each category. This ranking of the categories is reflected in each of the rows and columns of a fluctuation diagram if the variables are independent of each other. "Unusual" cells are therefore those that are — compared to their surroundings — too small or too large. One of the most obvious examples in Figure 5.11 is the high number of users in the UK giving *school* as their primary access to the web; relatively few users from Washington, D.C., The Netherlands, and Germany give *home* as their primary access to the web. For these countries, the primary access to the web is stated as *primarily work*.

Figure 5.12 shows the diagram for a χ^2 test of independence between *Country* and *Opinion on Censorship*. The cell sizes are proportional to the product of the marginal distributions. Red- and blue-coloured areas show the strength of deviation from independence. Blue-coloured areas indicate a positive residual: more cases were observed than would have been expected; red areas accordingly show negative residuals, where fewer cases were observed than would have been expected. The question asked in the survey was "Do you agree with the following statement: 'Certain information should not be published on the internet?'." The strongest disagreement came from Wisconsin, European countries, and Canada. The statement found most support in Connecticut, California, and Michigan.

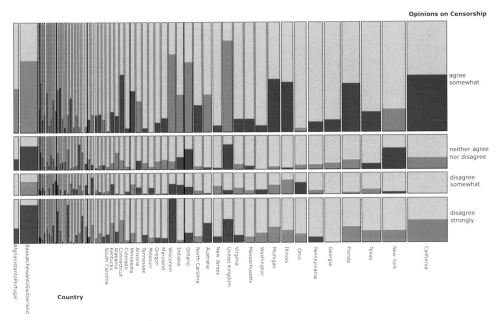

Fig. 5.12. *Diagram of a* χ^2 *test of independence between* Country *and* Opinion on Censorship *from the internet survey data. Red areas indicate negative residuals (appear dark gray); blue areas show positive residuals (appear black).*

5.4.4 Gray-shading

As well as different layouts, *gray-shading* of the bins can be used to emphasize a property or encode further information into the diagram. Figure 5.13 shows a mosaic plot in same binsize mode for the answers to the question "How did you hear about the survey?" There were nine possible answers: *Banner, Friend, Mailing List, Printed Media, Remembered, Search Engine, Usenet News, WWW Page,* and *Others,* yielding a total of 2^9 possible combinations. Figure 5.13 gives an overview of the choices made. Most users reported only one source, which results in lots of empty cells — more than 20% of all possible combinations are empty. To make empty cells visible but distinguish them from other cells, empty cells are marked by a small circle in the middle of the cell. Whenever a combination is empty during the construction of a mosaic plot, this cell is not split further. This means that higher-order combinations of empty cells are combined into one large cell in the final display, making the result easier to interpret. In Figure 5.13, the largest cluster of empty cells is the combination corresponding to the tile in the middle of the bottom row. No user had heard about the survey in the combination of printed media and a friend and remembered from a past experience but not seen the survey from the WWW site. In the same binsize representation, all bins carry the same weight visually. To emphasize differences in the number of counts, additional colouring is used. The gray-shading represents the actual cell size. Lighter colours mean higher numbers of users reporting. The largest cell corresponds to the choice "heard from WWW page" only.

By default, the relationship between cell size and colour is linear. Depending on the distribution of cell sizes, different transformations emphasize different aspects. In Figure 5.14, the following transfer function is used: Assume cell sizes x have been standardized to a range of $(0, 1)$ by $x \Rightarrow \dfrac{x - \min}{\max - \min}$. The transfer function $f_{\alpha, \beta}$ is defined as:

$$f_{\alpha, \beta}(x) = \begin{cases} \alpha^{1-\beta} \cdot x^{\beta} & 0 \le x \le \alpha; \\ 1 - (1 - \alpha)^{1-\beta} \cdot (1 - x)^{\beta}, & \alpha \le x \le 1, \end{cases}$$

resulting in a continuous and differentiable function with a fixpoint at α and $f'(\alpha) = \beta$. For skew data, as is used for the gray-shading shown in Figure 5.14, the default parameters $(\alpha, \beta) = (0.5, 1)$ yield a mapping that is predominantly black with a few instances of white in between for bins with very large numbers of cases (middle of Figure 5.14).

By switching from the linear mapping to a mapping with $(\alpha, \beta) = (0.8, 0.2)$, an S-shaped function results, which is able to differentiate be-

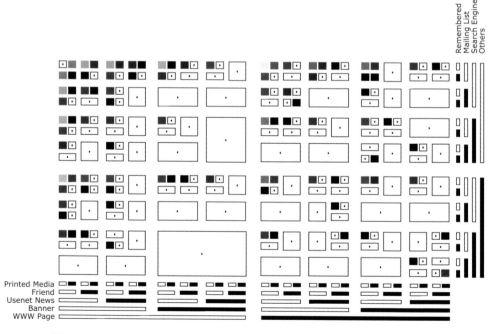

Fig. 5.13. *Mosaic Plots of all possibilities of "How you Heard about the Survey?" Gray-shading has been used to mark cell sizes. Empty cells are combined where possible and tagged by a circle. 112 of the possible 512 combinations are empty.*

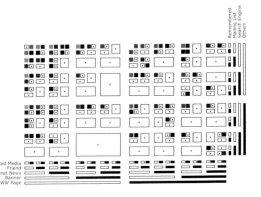
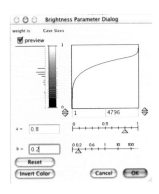

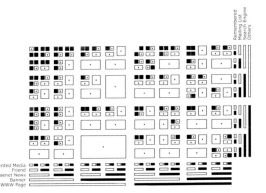
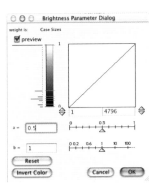

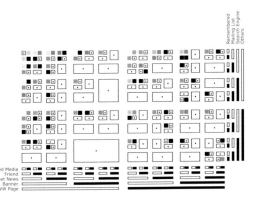
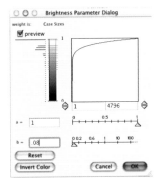

Fig. 5.14. *Dialog box for setting up the parameters of the brightness function. From top to bottom the parameters are* $\alpha = 0.8, \beta = 0.2$; $\alpha = 0.5, \beta = 1$ *(default), and* $\alpha = 1, \beta = 0.08$.

tween bins with small numbers of cases. Overall, the mosaic plot is lighter (top of Figure 5.14), introducing various shades of dark gray.

The plot on the bottom of Figure 5.14 is the lightest in colour, with the biggest spread of differentiation among bins with small numbers of cases. The mapping function has parameters $(\alpha, \beta) = (1, 0.08)$, i.e., the previous S-shape is replaced by a function with strictly positive curvature.

Gray-shading is discussed for binned plots in Section 6.2.3.

5.4.5 Rescaling Binsizes

Typically, the distributions of binsizes in mosaic diagrams of variables with high numbers of categories are very skew. A few cells dominate the diagram if the areas are directly proportional to cell sizes.

The focus can be brought back to smaller bins if they are artificially increased in size — up to a specified upper limit (ceiling-censored zooming). Figure 5.15 shows two fluctuation diagrams of the set of nine binary variables related to "How did you hear about the Survey?" Each of these variables corresponds to a checkmark choice in the survey. The choices were given as *Followed link from another Web page; Followed banner from another Web page; Found link using a search engine; Saw postings to WWW related newsgroups; Received email from www-surveying mailing list; Was told URL by friend; Read about it in a newspaper/magazine; Remembered to participate from last survey;* and *Other sources.* On average, 1.09 sources were checked by each person participating in the survey, so only very few combinations of the nine sources out of the 512 possible ones will have a large number of cases in them. This becomes apparent in the default fluctuation diagram on the left side of Figure 5.15, where very few rectangles are visible. The largest bin corresponds to *Followed link from another Web page* as the only source of knowledge about the survey, comprising 47.5% of all the data.

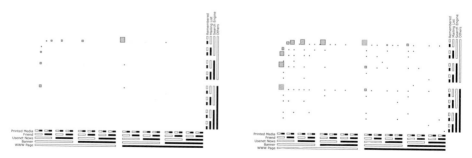

Fig. 5.15. *Fluctuation diagrams of all 512 possibilities of "How did you hear about the Survey?" In the diagram on the right, small bins are over-emphasized. Bins that have already reached the "ceiling" limit are marked by a red frame.*

On the right side of Figure 5.15, the bins have been made bigger while
keeping the aspect ratio fixed. The two largest bins have reached the maximum space limit and are marked by red frames. In all four quadrants of
the fluctuation diagram, additional bins appear — most of them fall in the
upper left quadrant, very few in the lower right. At the very bottom right
of the plot on the right there is a small group of people who answered yes
to all nine options (querying reveals it was 27 of them).

Instead of focusing on bins with small numbers of cases, the interest
might be on the largest bins while treating the smaller bins as noise.
This suggests a different approach to visualize cell sizes by starting from
a same binsize display and blending out cells with small cell sizes (floor-
censored zooming).

The importance of zooming functionality for large datasets is also discussed in Section 4.4.3.

5.4.6 Rankings

In the case of extremely skew bin distributions, as with the data from
"How did you hear about the Survey?" it is worth just concentrating on
the ranking of each combination. Instead of the true cell sizes, the fluctuation diagram of Figure 5.16 shows areas that are proportional to the
cells' order, starting with an order value of 0 for empty cells up to 512 for
the combination with the most entries: "heard from WWW page only."

5.5 Summary

Large datasets imply not only large numbers of cases but also large numbers of categories in each variable. Large numbers of cases do not affect area-based diagrams, such as barcharts or mosaic plots, very much.
With increasing numbers of variable categories, though, even the simplest tasks such as comparisons of bin heights in barcharts become more
difficult. Interactive tools such as reordering bins, grouping of bins or
rescaling the axes have been introduced to overcome these problems.

It is important to keep track of small numbers of data points and at
the same time provide overviews of the data. With high numbers of categories, the one-dimensional distribution of variables tends to become very
skew — which makes scaling awkward and comparisons virtually impossible: a few categories usually dominate the whole display. Methods to al-
leviate this problem are a change of the axes or weighted displays, where
the weighting variable can, for example, represent just the rank of a category.

For higher-dimensional displays such as the mosaic plot, the problems
due to high numbers of categories are multiplied because they arise in
each dimension. Additionally, higher-order effects arise: the data tables

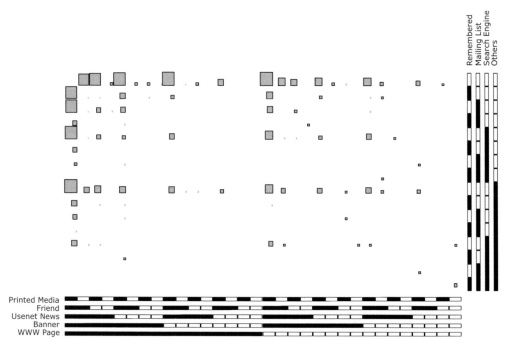

Fig. 5.16. *Fluctuation diagram for "How did you hear about the Survey?" The area of each cell corresponds to the cell's rank.*

tend to have many cells with zero counts and the corresponding mosaic plot therefore has many empty bins. The skewness of the distribution affects the display even more. Grid-based variations of mosiac plots, such as fluctuation diagrams, or even a simple approach such as same bin-size diagrams are very helpful for a first representation of the data. With additional sorting and grouping techniques, interesting patterns in the data often take the form of cells that are locally "too big" or "too small". Within overview displays, it is possible to identify data anomalies and small groups with distinct behaviours.

6

Rotating Plots

Dianne Cook and Leslie Miller

Love can tell and love alone, whence the million stars are strewn.

Robert Bridges, *My Delight and Thy Delight*

6.1 Introduction

This chapter discusses using preprocessing of multivariate real-valued data into projections displayed as scatterplots converted to images and movies to provide graphical displays for large quantities of data. The views are augmented with interaction, so that the user can paint areas and have small subsets of the full dataset overlaid as real-time graphics. Looking up the full dataset is achieved by an indexing of the projections that is created in the data preprocessing. An application of the methods to multivariate spatio-temporal data, seasonal metric data from the United States Geological Survey, is discussed.

Large spatio-temporal datasets are being compiled to study global climate and environmental conditions across the globe. A lot of data is being collected with the support of the United States government and is being made available to the general public: see, for example, http://www.pmel.noaa.gov/tao, http://www.usgs.gov/satellitedata.html, http://www.usgs.gov/digitaldata.html and http://www.ucar.edu/fac_data/data-resources.html. One of the hurdles to using existing graphics tools on these data is the size. For real-time visualization, large data present some challenges:

1. The views need to be created rapidly enough to be displayed in fractions of a second, especially for dynamic graphics;
2. A user needs to be able to interact with the views and have the response occur in fractions of a second;
3. The screen real estate is limited and with large data many cases could map to the same pixel;

4. Case-reduction methods for multivariate real-valued data, while sufficient for mean and variance analysis, are inadequate for visualization where the tasks are to find anomalies, rare events, or uncommon relationships.

The objective is to produce visual tools that provide displays swiftly, allow the user to interact with the view, and maximize the amount of data represented within the limits of screen resolution. For this chapter, the goal is to demonstrate that this is achievable for a million cases of multivariate data.

As early as 1990, McDonald, at the University of Washington, demonstrated software for rotating more than a million points, but he never published these research findings. He achieved rotation of a million points by preprocessing the scatterplots of the projections of 3-D to 2-D into images. The rotation started slowly and discontinuously: first one view, then the next, then the next, stepwise as the images for each projection from 3-D to 2-D were created, and then it would scream along at breakneck pace once the full set of images for 360° rotation was created. He could also brush in a scatterplot of more than a million points linked to other scatterplots. McDonald achieved this by controlling the plot updates — updates were only made when the mouse button was released. The scatterplots were also displayed at pixel resolution, gray scale images, and colour was overlaid in layers for each brush action. The software was used to examine data generated from remote-sensing instruments, that is, images representing a spectral band. This type of data typically comes on the order of a million points, $1,024 \times 1,024 = 1,048,576$, and multiple bands (dimensions).

The research reported in this chapter revisits McDonald's work, adding methods and software for visualizing large, multivariate space-time dependent data. It is organized as follows. The rest of this section introduces notation for the data, describes visual methods commonly used for small datasets, and provides more information on related work. Section 6.2 describes the approach to scaling methods to a million cases. Section 6.3 overviews the software, called Limn, developed to test the approach, and Section 6.4 applies the methodology to seasonal metric data from the United States Geological Survey (USGS).

6.1.1 Type of Data

The data are real-valued multivariate data having at least a million cases (instances, examples), but probably fewer than 15 variables (features, attributes). In some examples, there may be associated class information, such as an estimate of land use, and there may be associated time and spatial information.

In mathematical notation, the data are expected to have the form

$$X = \begin{bmatrix} x_{11} & x_{12} & \cdots & x_{1p} \\ x_{21} & x_{22} & \cdots & x_{2p} \\ \vdots & \vdots & \ddots & \vdots \\ x_{n1} & x_{n2} & \cdots & x_{np} \end{bmatrix}_{n \times p} = \begin{bmatrix} \text{tuple}_1 \\ \text{tuple}_2 \\ \vdots \\ \text{tuple}_n \end{bmatrix},$$

where n is the number of cases, and p is the number of variables. The associated class vector may be denoted as $C = \begin{bmatrix} c_1 & c_2 & \cdots & c_n \end{bmatrix}'$, the time context as $T = \begin{bmatrix} t_1 & t_2 & \cdots & t_n \end{bmatrix}'$ and the spatial context as $S = \begin{bmatrix} s_{11} & s_{12} \\ s_{21} & s_{22} \\ \vdots & \vdots \\ s_{n1} & s_{n2} \end{bmatrix}$.

There may be missing values in the data.

A data projection, as used to generate a scatterplot, for example, can be denoted as

$$Y = XA = \begin{bmatrix} x_{11} & x_{12} & \cdots & x_{1p} \\ x_{21} & x_{22} & \cdots & x_{2p} \\ \vdots & \vdots & \ddots & \vdots \\ x_{n1} & x_{n2} & \cdots & x_{np} \end{bmatrix}_{n \times p} \begin{bmatrix} a_{11} & a_{12} \\ a_{21} & a_{22} \\ \vdots & \vdots \\ a_{p1} & a_{p2} \end{bmatrix}_{p \times 2} = \begin{bmatrix} y_{11} & y_{12} \\ y_{21} & y_{22} \\ \vdots & \vdots \\ y_{n1} & y_{n2} \end{bmatrix}_{n \times 2}$$

where A is an orthonormal matrix, that is the columns are vectors of length 1 and each is orthogonal to all other columns.

6.1.2 Visual Methods for Continuous Variables

There is a lot of choice in graphical methods for displaying aspects of multidimensional continuous data: tours (Asimov; 1985; Cook et al.; 1995; Wegman; 1991); scatterplot matrices (Carr; 1985; Becker and Cleveland; 1987); parallel coordinate plots (Inselberg; 1985; Wegman; 1990); and conditional plots (Becker et al.; 1996) (cf. Section 2.3.2). Scatterplots of two real-valued variables underlie several of the above methods, and they can be thought of as projections, projecting the data from p to 2 dimensions.

Linking between multiple plots provides ways of relating information, giving insight into high-dimensional structure in the data (see also Section 4.3.2). A commonly available approach is linked brushing, where highlighting features in one rendering highlights the corresponding features in the others. Linking marginal views allows the viewer to explore conditional dependencies amongst variables. For example, Figure 6.1 shows plots from the olive oil dataset. In Region 2, linoleic acid is relatively high and oleic acid is relatively low (middle plot). A combination of variables 6, 7, 9, 10 (oleic, linoleic, arachidic, eicosenooic) separates Region 2 points from all other points (right plot).

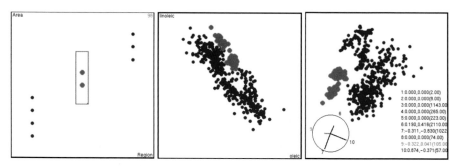

Fig. 6.1. *Linking between several views: (left) brushing on class value, (middle) scatterplot of two variables, (right) tour projection.*

Linking is well-documented in Cleveland and McGill (1988) and Buja and Tukey (1991).

There are some articles in the collection Card et al. (1999) and some brief treatments in Spence (2000). Numerous software environments provide these types of tools: for example, GGobi (http://www.ggobi.org), Mondrian (http://www.theusRus.de/Mondrian), MANET (http://www.rosuda.org/Manet/), DataDesk (http://www.datadesk.com), JMP (http://www.jmpdiscovery.com), Orca (http://software.biostat.washington.edu/orca/), XLispStat (Tierney; 1991), Spotfire (http://www.spotfire.com), XmdvTool (http://davis.wpi.edu/~xmdv/). For the spatial data used in this chapter, the best description of interactive graphical methods can be read in Unwin et al. (1990) and Haslett et al. (1991).

6.1.3 Scaling Up Multiple Views for Larger Datasets

Chapter 3 introduced several general techniques to scale up data displays for large datasets. Dan Carr has published several articles on large data visualization for scatterplots, using mostly static binned views (Carr; 1991). Wegman et al. (1998) developed a "pixel tour" for displaying (multivariate spatial-context) spectral data, in which sequences of one-dimensional projections of the multiple spectral bands are displayed as gray scale images. This is very useful for detecting major events in the spatial domain, such as mines in a minefield.

6.2 Beginning to Work with a Million Cases

6.2.1 What Happens in GGobi, a Real-time System?

The software GGobi (Swayne et al.; 2003) was developed for graphical projection pursuit in multivariate data. As with all research software it was used first for small datasets, and now interest lies in how well it can cope with large datasets and what additional developments may need to be introduced.

It is possible to get a million cases of multivariate data into the software GGobi (at least on a Linux system). The startup time is long: it took about 7 minutes. This is probably due to the file-reading routines that read the data a line at a time and block further computation until the entire file is read. Once the data have been imported, there are two problems: overplotting and slow response time. A few modifications to the usual user behaviour can accommodate exploring this many cases:

- Use a pixel-sized glyph. The smaller glyph reduces the overplotting that obscures the data distribution of the large number of points.
- Keep the brush off while positioning it in a plot, and turn it on when ready to paint a subset of the data. This is equivalent to McDonald's method for controlling the plot updates. The plot is not updated continuously when the brush is moving. It will be updated only when the brush is turned on again.
- Keep the tour paused because the re-plotting is too slow to give smooth continuous motion. Use the scramble button, which offers random projections of the data.
- Avoid methods which involve real-time binning of the data, such as ASH (see also Section 2.3.1). The binning code in this implementation is too inefficient to compute in real-time with this size of dataset. Methods like α-blending — putting the load onto the graphics hardware — might be more efficient (see Section 3.5).
- Keep the number of open displays small. Fewer open displays means less plot updating.

6.2.2 Reducing the Number of Cases

How can the number of cases in a multivariate dataset be reduced? One natural approach is to bin the data in the multivariate space (p-dimensional), either by using rectangular boxes or by bins induced by a cluster algorithm. Figure 6.2 (left and middle) illustrates these approaches. An alternative approach is to bin the data in the space that will be plotted, e.g., 2-D for a scatterplot or a projection of the data (Figure 6.2 right).

There has been considerable research into case reduction in the multivariate space. Techniques that are commonly in use are

- **Binning**: cases are gridded into cells (Figure 6.2 left). The points in the cell are represented by the bin center or by the average of the points in the cell, and a count of how many cases are in the cell. Using the average value better reflects the data distribution within the

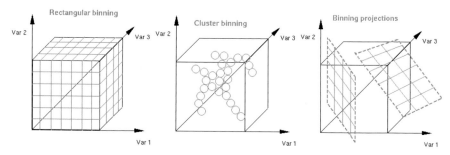

Fig. 6.2. *Illustration of three different types of binning for 3-D data: (left) rectangular binning in the multivariate space, (middle) binning using a clustering algorithm, (right) binning of the projected data.*

cell when the cases are non-uniformly distributed. Binning is inefficient when there are more than several variables. The number of bins needed is exponential in the number of variables. If each variable is binned into 10 cells, there are 10^2 cells for 2 variables and 10^{10} cells for 10 variables. It also creates visual artifacts, grid patterns, which can distract the user from seeing the structure of the data. There is also serious potential for loss of resolution with binning.

- **Clustering:** a cluster algorithm is run, usually k-means with k set to be very large (Figure 6.2 middle). Then cases are binned to their closest cluster mean, and each case has an associated weight representing the count of the number of cases in the bin. (See, for example, Posse (2001).)
- **Data squashing:** data are binned as a first step, then statistics are computed on each bin, e.g. mean, variance/covariance. These statistics are then used to generate a new sample in each bin to construct a reduced case dataset with similar statistical properties to the original one (DuMouchel et al.; 1999).

With any of these methods the resulting dataset is of the form:

$$X^* = \begin{bmatrix} x_{11} & x_{12} & \cdots & x_{1p} \\ x_{21} & x_{22} & \cdots & x_{2p} \\ \vdots & \vdots & \ddots & \vdots \\ x_{n^*1} & x_{n^*2} & \cdots & x_{n^*p} \end{bmatrix}_{n^* \times p} \qquad W = \begin{bmatrix} w_1 \\ w_2 \\ \vdots \\ w_{n^*} \end{bmatrix}$$

where $n^* < n$ and each case has an associated count or weight, $w_i, i = 1, \ldots, n^*$, corresponding to the relative representation of this combination of values in the full dataset. These techniques were explored in Han (2000), but did not give satisfactory results. Binning may be appropriate for numerical analysis, when a mean or variance analysis is the focus; but when finding anomalies and deviations from a trend, the artificial cluster structure induced by the binning confounds the task (Figure 6.3).

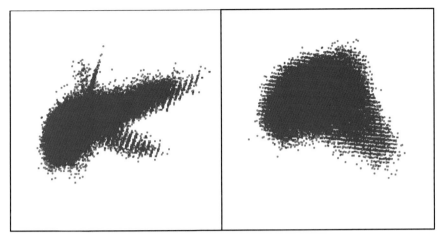

Fig. 6.3. *Two views of five-dimensional data that have been binned to* 30(!) *cells per variable. There is artificial clustering simply due to the binning.*

Because of the shortcomings of techniques to reduce the size of the multivariate space, the approach concentrates on binning induced by the screen real estate for each projection (Figure 6.2 right). The binning will be different for each projection, but this form of binning also induces a reduced size structure on the data. The binning induced by the screen real estate essentially results in case reduction for *each projection*: screen horizontal position, S_x, and screen vertical position, S_y, and a count or weight corresponding to the number of points plotted at the pixel, $W_{S_x S_y}$. (An ID list, $P_{x,y,j}$, $j = 1, \ldots, \#$number of points at pixel S_x, S_y, is also defined to refer to the row value of the original data matrix.) For example, if $n = 1,000,000$ and the display window is 200×200 then there are at most $40,000$ plotting positions, so that the data are reduced from $n = 1,000,000$ to $n^* \leq 40,000$ in each projection. Multiple projections of the data are created. This structure is utilized to build a fast lookup system for linking points between plots. Figure 6.4 illustrates the data projection into a window on screen: the raw data are standardized and projected and then scaled again to be represented at a screen position.

6.2.3 Density Estimation

The binned data projection is represented as a density plot where the bin count for each pixel is coded as a gray scale value. The transfer function used is $gray\ scale = 255 \times \left(\dfrac{count}{maxcount}\right)^p$. Figure 6.5 shows the function for several different values of the power, p. The reason for using a power transfer function is that the human eye cannot readily discern differences between adjacent gray values on a linear scale. This transfer function

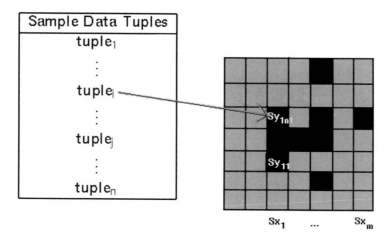

Fig. 6.4. *Data are projected into pixel locations in a screen on the window. Black represents a pixel that contains at least one data point, gray represents no data at the pixel. The index is built only on pixels containing data.*

was chosen after examining the techniques discussed in Huang et al. (1997) and the lecture notes for Purdue University course EE637 Digital Image Processing I `http://dynamo.ecn.purdue.edu/~bouman/ee637/`. (Another alternative, allowing more flexible transfer functions, is described in Section 5.4.4 and illustrated in Figure 5.14.)

The power transfer function effectively reveals more perceptible resolution of the data density in regions of high density, while also rendering the extreme points in dark ink. This can be seen in Figure 6.6: the top row of density plots uses a linear transfer function, and the effect of "lightening" up the high–density region as the power value decreases can be seen going down the page. The function renders low density as black and high density as white — similar to what you might physically see flying over snow-capped mountains. Zero count is also rendered as white. In general, this is not a problem in tour views especially as the high–density regions (white) are separated from the zero count (white) by low–density pixels (black). And the dark ink for the low–density regions draws the eye to the extreme anomalies — what you want to detect using graphics.

For multiple projections it may be useful to use the maximum bin count over all the projections, rather than the maximum bin count in a single image, to scale the bin counts. This will be particularly important for maintaining smoothness of gray scale from image to image in a tour movie sequence. The underlying concept of multiple views and common scaling of plots is discussed in Section 4.4.3.

Two problems arise with density plots: both are evident in the examples in Figure 6.6. When the bin count is small and the maximum count

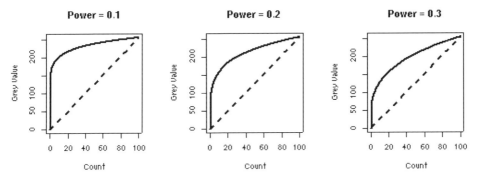

Fig. 6.5. *Transfer functions used to convert count to a gray scale: three different power functions. The dotted lines represent linear transfer functions.*

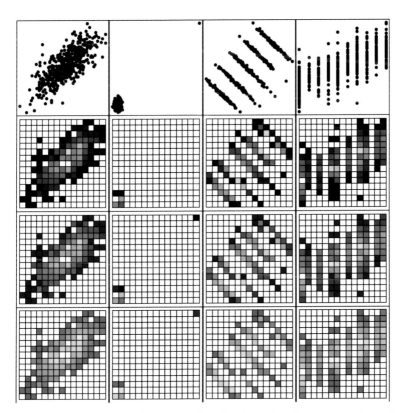

Fig. 6.6. *Four very different bivariate distributions that have approximately the same correlation, and their binned views using different power transfer functions: (second top) linear, (second bottom) power = 0.6, (bottom) power = 0.3.*

is large, the bin may get rendered as a zero count cell. This happens with the linear transfer function for the example with a single extreme value (second left column). The other transfer functions correctly display the cell as having a non-zero count. The second problem occurs when a variable takes integer values with the total number being less than the screen real estate size (right column). The result is that strips of white space are visible periodically, a very undesirable effect. These problems are similar to those encountered when using histogram displays of single variables. The first problem can be alleviated by interactively changing the parameters of the transfer function, but the second cannot because the plot size is set, the data binned, and then the plots produced, meaning that it is not possible to interactively change the binning.

6.2.4 Screen Real Estate Indexing

A three-level index is created for each stored data projection. In its simplest format, the index consists of three files.

- The Sx index file stores S_x coordinates of the non-empty pixels by increasing order and pointers to the starting position of the corresponding Sy-blocks in the Pixel index file.
- The Pixel index file stores S_y coordinates of the non-empty bins and pointers to the starting position of the corresponding ID list in the IDList index file. Each block of S_y coordinates is ordered in ascending order.
- The IDList index file stores ID numbers that point to the raw data tuples that are projected to a point in this plot image.

    With a small dataset, a full ID list can be stored for each non-empty bin of a binned projection. But for large datasets, with millions of tuples, space considerations make it impractical to maintain such a full index. To make use of the index for large datasets, it is necessary to reduce the size of the ID lists while maintaining as much information as possible. Two approaches have been used for this in the software, Limn. First, the complete dataset is used to calculate the count, mean, minimum, and maximum of the set of tuples that plot to a given point. The statistics are stored in a fourth file that is linked back to the index files, allowing Limn users access to the statistics during execution. Second, sampling of the complete set of tuples that plot to a bin is used to provide access to a reasonable set of tuples that represent the point. In the current prototype random sampling is being used. Future development plans call for other methods of sampling to be evaluated as Limn testing continues. A block diagram of the current index structure is shown in Figure 6.7.

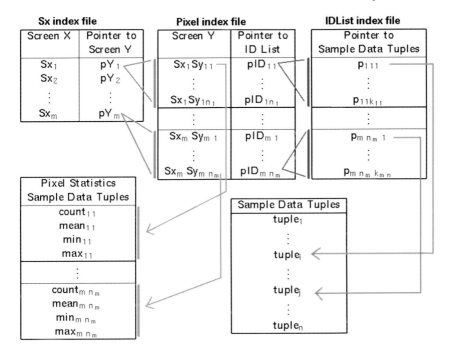

Fig. 6.7. *Diagram illustrating the three layer index files structure.*

6.3 Software System

The Limn system approaches the performance issues associated with real time viewing of large datasets by preprocessing the data when possible. This preprocessing can include both indexing the data for faster searches and generating sets of images of the data to provide an animated backdrop. Dynamic graphics can then be generated in front of the backdrop images. The system is designed as a pluggable framework that allows new modules to be easily incorporated into the data preprocessing and viewing stages. Current modules provide support for reading in data from multiple sources, creating images and their indices to the original data for a scatterplot matrix, and viewing programs to display the images. A tour animation sequence can be built as a quicktime movie or assembled from a collection of image files. The View classes can be used to view the images and interact with them.

Functionally, it reduces to two main activities: preprocessing the data and viewing the results. Figure 6.8 outlines the two structures.

• The Limn preprocessing program controls processing the data into projections and indexing the projections. If an animation sequence is to be the result, then a quicktime movie is created. A properties file is used

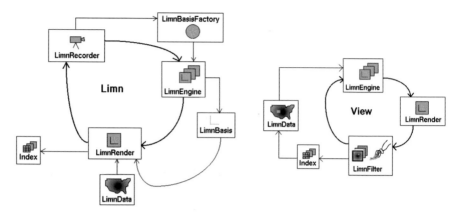

Fig. 6.8. *Diagram illustrating the preprocessing (left) and viewing (right) struc-*
tures.

to specify animation creation and appearance attributes. The program
creates the tour engine, rendering and recording type objects that are
specified in the properties file and starts the animation cycle. If a scatterplot matrix is to be the result, the Limn program stops at simply
building the index, and leaves the image rendering to the View part of
the program. For large datasets, the indexing routine will keep only a
sample of indices for the points at each pixel.

- The viewing part of Limn will display a scatterplot matrix or a tour
  animation that allows the user to control the speed and direction of
  the playback. Additionally, users can interact with the display while
  an animation is running. By brushing an area of a plot, data points
  will be selected and the viewer will draw graphics around each point
  to highlight it. Because the highlighted graphics are drawn on top of
  the scatterplot image, performance will begin to lag as the subset size
  gets larger.

Users interact with a scatterplot image by brushing areas on the plot
and have the corresponding subset overlaid on other plots as real-time
graphics. When a bounding box is brushed on a scatterplot by a user,
the bounding box determines the range of S_x and S_y. The Sx index file
makes use of the S_x range to determine the relevant S_y lists in the Pixel
index file. The Pixel index file then matches the S_y range only to those S_y
values that satisfied the S_x range criteria to determine the ID lists in the
sample IDList index file. The tuples from the sample list for each relevant
scatterplot point are retrieved. The resulting data can then be linked to
the other scatterplots of interest to the user. The result is that brushing
can be applied to much larger datasets.

6.4 Application

6.4.1 Data Description

The US Geological Survey's (USGS) EROS Data Center produces bi-weekly composite images of the Normalized Difference Vegetation Index (NDVI). NDVI is a spectral index based on the surface reflectivity esti-mated by the first two spectral bands of the Advanced Very High Resolu-tion Radiometer (AVHRR) which operates on several National Oceano-graphic and Atmospheric Administration (NOAA) satellite platforms. This combination is used as an index to measure the health and density of plant biomass.

The datasets include estimates for the start, end, and peak of the growing season for each square kilometer of the US, as well as measures for biomass at those times and for net productivity over the growing sea-son. The value of this dataset is that it can be used to classify land cover by how it functions, not just by how it appears.

The seasonal metrics used in this study are SoST (Time of the start of the growing season), SoSN (NDVI at the start of the growing season), EoST (Time at the end of the growing season), EoSN (NDVI at the end of the growing season), TotalNDVI (Integrated NDVI over the season — an indicator for net productivity). In addition, each spatial region is classi-fied into one of 159 EPA ecoregions. There are 15 classes of agricultural use (11/17 is corn/soybeans, 56 is grasslands, 94 is deciduous forest, 139 is coniferous forest, 155 is barren ...) at each spatial location, and measure-ments for several time periods. Data from 1989 are used for this example, which comes to a little over a million points.

The primary objective of the data collection is to monitor the health and density of the nation's biomass over the growing season. It is also of interest to examine land cover class in relation to seasonal metrics.

The data arrive as multi-band, 16-bit, binary flat files with each band corresponding to a different year and each file containing a single vari-able. Each band in this dataset can be thought of as an image map of a single variable for each year. Figure 6.9 demonstrates a three-year bi-nary file for Start of Season Time. The data input capabilities of Limn are limited to reading ASCII columns of text, where each column represents a variable and each row represents a single latitude–longitude pair for a single year. The data were preprocessed using other software to get it into a format that Limn could use.

6.4.2 Viewing a Tour of the Data

Figure 6.10 displays some snapshots from the tour animation. The top left display shows the full window with the user beginning to brush, and the right top shows the results of the brushing (green). The plots below

SOST

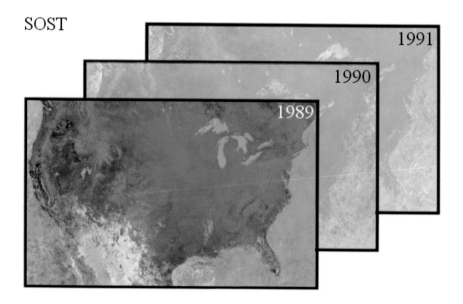

Fig. 6.9. *Representation of three–year binary file for the Start of Season Time.*

show a sequence of three tour projections, where the subset is displayed as real-time graphics on the movie images. What can be seen? In the high-density region, there is substantial multimodality which most likely corresponds to different land use types. Around this region, there are numerous singleton or low–density pockets. The brushed subset is one such pocket of points. As the tour progresses, this group spreads out, so it contains spatial locations that have similar extreme values in two variables but quite different values on other variables: that is, it is not clustered tightly in the five-dimensional space.

In the same way, many of these anomalies on the fringe of the scattercloud can be investigated. Small subsets are brushed and watched through the movie, then turned off, and another group brushed and watched. Could these be plots of land which are growing some unorthodox crop or could there be some kind of data quality problem?

6.4.3 Scatterplot Matrix

Figure 6.11 displays the scatterplot matrix of the class variable and the seasonal metric data. What can be seen from this?

In the figure, the classes corresponding to corn/soybean (green) and grasslands (pink) are brushed. There are substantial differences between these groups. One occurs in the SoSN and SoST plot (second row, right

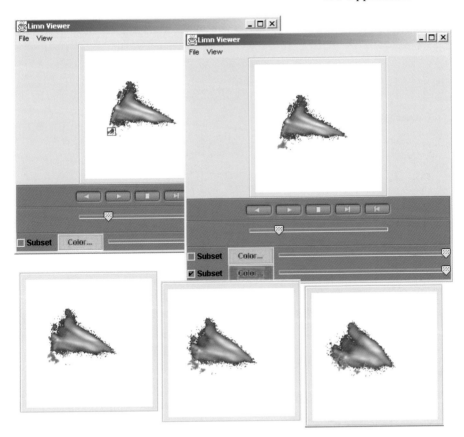

Fig. 6.10. *Sequence of tour views demonstrating brushing a subset and subsequent touring. The anomalous points on the fringe of the point cloud are the interesting ones to investigate. One such group has been brushed here. It spreads out and separates into smaller groups in other projections, so it is not a cluster in five-dimensional space but rather in perhaps 2 of the variables only.*

plot): grassland has relatively high SoSN (start of season NDVI), whereas corn/soybeans have a very diverse distribution in these two variables. There is also a difference in the EoSN, again grasslands have a fairly concentrated moderate value for EoSN.

Overlaid on the matrix is a snapshot of the user controls. When the user brushes in an image, a subset control tab is added to the control panel, and also a subset information frame is added to the information panel. The user can then control whether a sample from each pixel location, or the full data, or statistics such as the mean are plotted over the images. The user can also choose a different colour to represent the subset.

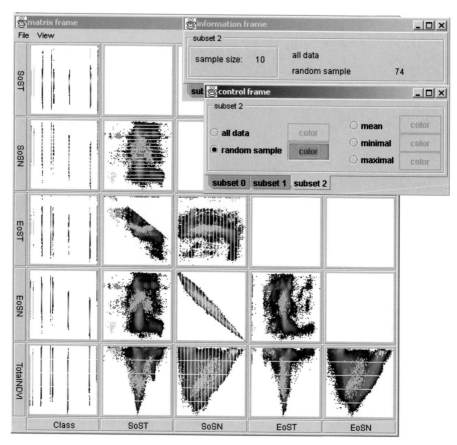

Fig. 6.11. *Scatterplot matrix of class and seasonal metric variables.*

6.5 Current and Future Developments

This chapter has described building graphics on a dataset containing a million cases of five-dimensional measurements. Further developments include overlaying multiple movies and this has also been implemented (although that work is not reported on in this chapter). The next stages are to think of where the approach needs improving, of where the software needs developing, and to test the tools with other applications.

6.5.1 Improving the Methods

With large datasets, there are too many points represented at each pixel to store and overlay on other projections, so it is planned to explore sampling schemes for the pixel populations. Currently, a threshold number is used, and all pointers to the raw data are kept up to a fixed sample size.

Thus, pixels where few points are represented maintain the full list of cases, but where there is high density a fixed number of randomly sampled points is kept.

6.5.2 Software

Large datasets are often stored as specially structured binary files, for example, netCDF (`www.unidata.ucar.edu/packages/netcdf/`, hdf (`hdf.ncsa.uiuc.edu/`), or some type of image format. Currently, the data are extracted out of this format with other image manipulation tools and stored in ASCII files. This is laborious and space-consuming. It makes more sense to develop data parsers to work with the binary formats.

The datasets will still be too large to operate the index preprocessing on the entire dataset simultaneously, but it could be run on different pieces of the data in parallel. We would like to explore this.

6.5.3 How Might These Tools Be Used?

As trivial as it might sound, one of the most likely uses of these tools is data cleaning. This is an important part of any data analysis. Through generating the example application in this chapter, several serious problems with the data were uncovered.

Exploring multivariate data graphically reveals many interesting features in a dataset as the graphics in this chapter have shown. In future, it is intended to use these tools to examine large datasets in a multivariate manner, especially global weather patterns and anomalies.

7

Multivariate Continuous Data — Parallel Coordinates

Rida Moustafa and Ed Wegman

As nearer marched the million feet of columns closing in.

Thomas Hardy, *Leipzig*

7.1 Introduction

Wegman (1995) has suggested that there are two relatively critical thresholds in Data Mining. The first occurs around 10^6 to 10^7 observations. At this scale, visualizing individual observations, computation of $O(n^2)$ algorithms, and data transfer over standard or even high-speed Ethernet becomes problematic. Visualizing larger datasets would seem to require resorting to visualizing density estimates or other summary methodology. A second threshold appears when tertiary storage becomes necessary. In 1995, this occurred around 10^{12} bytes, although only a few years later terabyte storage on hard drives in a simple machine became feasible. Nonetheless, it is inevitable that data will ultimately be sufficiently massive that tertiary storage mechanisms (magnetic tape silos) will be necessary. In this setting, access speed is decreased by orders of magnitude, and interactive data analysis is not possible.

For categorical data (or data forced into a categorical framework), seeing a million observations is slightly less problematic than seeing a million, high-dimensional numerical observations. The mosaic plot, described in Chapter 5, provides an excellent tool in the categorical framework. Computer science-based data miners have resorted to the so-called data cube, which is essentially a multivariate frequency table with continuous data quantized into a relatively small number of bins. While not so easily visualized in dimensions larger than three, it does provide a method that can relatively easily handle 10^6 or more observations.

The combination of parallel coordinates taken together with saturation brushing has been demonstrated to handle at least 250,000 observations in eight dimensions or more (Wegman; 2003). This latter paper also indicates that single categorical variables can be handled via parallel coordinate displays using colour brushing to separate categories. All of these techniques are effective for relatively large datasets because the algorithms involved tend to be of $O(n)$ complexity.

Parallel coordinates were introduced by Inselberg (1985), and he has continued much of their recent mathematical development. Wegman (1990) was the first refereed paper suggesting the use of parallel coordinates as a method for exploratory data analysis and data visualization. In the approximately 20 years since the formal introduction of parallel coordinates into the refereed literature, the parallel coordinate idea has become universally recognized as an effective method for dealing with multidimensional, numerical data.

As parallel coordinate plots have received much attention in the field of information visualization, various implementations are around. They can roughly be classified into two groups. The first group comprises the tools primarily implementing parallel coordinate plot functionality as, e.g., ExploreN (Carr et al.; 1997), CASSATT (Unwin et al.; 2003), or Xmd-vTool (Rundensteiner et al.; 2002). The second group includes tools that offer a far wider set of graphical tools and thus are not necessarily focused on parallel coordinate plots like GGobi (Swayne et al.; 2003), Mondrian (Theus; 2002b), or iPlots (Urbanek and Theus; 2003). There are certainly more implementations; some of them are only commercially available, others are extremely subject specific.

7.2 Interpolations and Inner Products

Consider a set of multivariate observations of the form $\boldsymbol{\xi}_i = (x_{i1}, \ldots, x_{id})$, where $i = 1, \ldots, n$ is the index for the data set size and $j = 1, \ldots, d$ is the index for the dimension. Perhaps the earliest multivariate visualization tool was the Andrews plot (Andrews; 1972), which is given by

$$y_{i\theta} = x_{i1}/\sqrt{2} + x_{i2}\sin(\theta) + x_{i3}\cos(\theta) + x_{i4}\sin(2\theta) + x_{i5}\cos(2\theta) + \cdots .$$

For sake of definiteness, assume d is odd. Then consider the vector

$$\boldsymbol{n}(\theta) = \left(\frac{1}{\sqrt{2}}, \sin(\theta), \cos(\theta), \ldots \sin\left(\frac{d-1}{2}\theta\right), \cos\left(\frac{d-1}{2}\theta\right) \right).$$

Then, $y_{i\theta} = < \boldsymbol{\xi}_i, \boldsymbol{n}(\theta) >$, the basic transformation for the Andrews plot, is an inner product of an appropriate vector with each observation, which, of course, is also a vector. If d is even, a simple modification of $\boldsymbol{n}(\theta)$ is appropriate.

In a similar manner, the Grand Tour is constructed by first forming a generalized rotation matrix $Q(\phi_{12\theta}, \phi_{13\theta}, \ldots, \phi_{d-1,d,\theta})$ and forming a rotated coordinate system with unit vectors

$$n_j(\theta) = e_j \times Q(\phi_{12\theta}, \phi_{13\theta}, \ldots, \phi_{d-1,d,\theta}), \ j = 1, \ldots, d$$

where e_j are the orthogonal basis vectors for the original coordinate system and $\times$ refers to matrix multiplication. Then, $y_{ij\theta} = < \xi_{i,}\ n_j(\theta) >$ is the projection of the ith observation into the jth rotated coordinate axis at time θ. Thus the Grand Tour can be conceived of as simply an appropriate set of interpolations using inner products. This is highly suggestive of a perspective for viewing parallel coordinates and also raises the possibility of additional suggestive interpolations. For a general description of the Grand Tour see Section 2.3.2 and Buja et al. (1996). A more detailed discussion of the framework in the context of parallel coordinates can be found in Wegman and Carr (1993) or Wegman and Solka (2002).

The projective geometry interpretation of parallel coordinates is well known. See Dimsdale (1984), Inselberg (1985), and Wegman (1990). One major feature is that points that are collinear in a two-dimensional Cartesian plot map into lines that are coincident in a two-dimensional parallel coordinate plot. If the slope of the line forming the points in Cartesian space is positive, the intersection of the corresponding lines in parallel coordinate space occurs outside of the space between the two parallel coordinates. Indeed, if the slope is $+1$ the lines in parallel coordinate space are parallel and the intersection occurs at ∞. Visually, this is not easy to diagnose. Conversely, if the slope of the line formed by the points in Cartesian space is negative, the intersection of the corresponding lines in parallel coordinate space occurs between the two parallel coordinate axes and, thus, linear structure is easy to diagnose visually. Wegman has called this the crossover effect. The crossover effect represents an easy way to reveal linear or near-linear structure in one or more dimensions. This is the motivation for additional interpolation methods.

7.3 Generalized Parallel Coordinate Geometry

The basic idea underlying parallel coordinates is to visualize hyperdimensional points in a sequence of parallel axes. The point $\xi = (x_1, x_2, \ldots, x_m)$ is visualized in parallel coordinates as a broken line (polyline) connecting the points $x_1, x_2, \ldots, x_m$ in m vertical or horizontal axes as shown in Figure 7.1. Although some have a slight preference for horizontal axes because of aspect ratio considerations on most computer screens, the orientation of the axes is essentially immaterial.

Consider for the moment the two-dimensional case and consider the following map: $T(\xi), \xi \in \mathcal{R}^2$, which maps a point $\xi = (x_1, x_2)$ into a line.

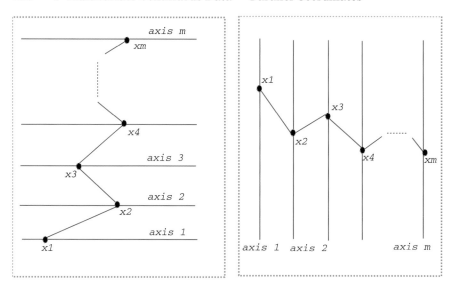

Horizontal Parallel coordinates   *Vertical Parallel coordinates*

Fig. 7.1. *An m-dimensional point visualized in parallel coordinates.*

This mapping can be constructed based on the transformation of the original space into a parameter space as

$$T : \xi \longrightarrow (t,\ g(t)),\quad 0 \le t \le 1,$$

where $g(t) = (1-t)x_1 + tx_2$. This transformation is a convex relationship between the first variable and the second. Furthermore, note that $g(t)$ is in the form of a Lagrange linear interpolation of degree one. Suppose the vector $n(t) = (1-t,t)$ is now considered and let $\xi_1 = (x_1, x_2)$. Then $T(\xi_1) = < \xi_1, n(t) >$ represents the parallel coordinate transformation as an inner product form. Indeed, if in general $\xi_j = (x_j, x_{j+1})$, then the mapping for a general multivariate vector $x = (x_1, x_2, \ldots, x_d)$ is represented by

$$T_j(\xi_j) =< \xi_j, n(t) >,\ 0 \le t \le 1,$$

where T_j is the interpolation between the jth and the $(j+1)$ parallel axes. Notice that the values $n(0) = (1,0)$ and $n(1) = (0,1)$ form an orthonormal basis for $\mathcal{R}^2$. If this particular interpolation leads to the Cartesian points mapping into parallel lines, alternate interpolations may be more satisfactory. Let us relabel the interpolating function $n(t) = (1-t,t)$ as $n_1(t)$ and introduce the following interpolations:

$n_1(t) = (1 - t, t)$ with $0 \leq t \leq 1$, where $n_1(0) = (1,0)$, $n_1(1) = (0,1)$

$n_2(t) = (t - 1, t)$ with $0 \leq t \leq 1$, where $n_2(0) = (-1,0)$, $n_2(1) = (0,1)$

$n_3(t) = (1 - t, -t)$ with $0 \leq t \leq 1$, where $n_3(0) = (1,0)$, $n_3(1) = (0,-1)$

$n_4(t) = (t - 1, -t)$ with $0 \leq t \leq 1$, where $n_4(0) = (-1,0)$, $n_4(1) = (0,-1)$.

In all four cases, the values of $n_k(t)$ at 0 and 1 form an orthonormal basis for $\mathcal{R}^2$. n_2 has the effect of flipping the x_1 axis, n_3 has the effect of flipping the x_2 axis, and n_4 has the effect of flipping both axes. These are illustrated in Figure 7.2. Notice that cases with parallel lines with slope equal to $+1$ can be made to intersect by an appropriate substitution of the interpolating vector.

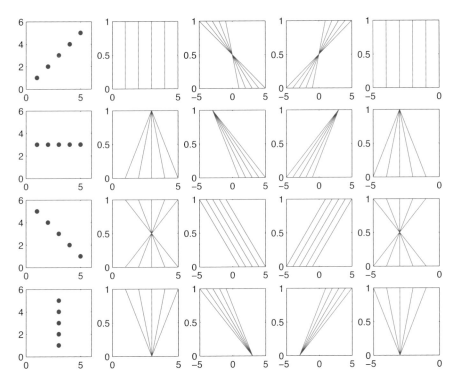

Fig. 7.2. *The data are in column 1. Their parallel coordinate plots with 4-possible normal vectors of $n(t)$ are in columns 2–5.*

A polar representation can also be made as follows:

$$\boldsymbol{n}_5(\theta) = (\cos(\theta),\, \sin(\theta))\text{ with }\theta \in [0, \pi/2] \text{ and } \boldsymbol{n}_5(0) = (1,0),\ \boldsymbol{n}_5(1) = (0,1)$$

$$\boldsymbol{n}_6(\theta) = (-\cos(\theta),\, \sin(\theta))\text{ with }\theta \in [0, \pi/2];\ \boldsymbol{n}_6(0) = (-1,0),\ \boldsymbol{n}_6(1) = (0,1)$$

$$\boldsymbol{n}_7(\theta) = (\cos(\theta),\, -\sin(\theta))\text{ with }\theta \in [0, \pi/2];\ \boldsymbol{n}_7(0) = (1,0),\ \boldsymbol{n}_7(1) = (0,-1)$$

$$\boldsymbol{n}_8(\theta) = (-\cos(\theta),\, -\sin(\theta))\text{ with }\theta \in [0, \pi/2];\ \boldsymbol{n}_8(0) = (-1,0),\ \boldsymbol{n}_8(1) = (0,-1).$$

Again, in all four cases the values of $\boldsymbol{n}_k(\theta)$ at 0 and $\pi/2$ form an orthonormal basis for $\mathcal{R}^2$ and have essentially the same flipping effect as the previous four interpolations. The effect of the sinusoidal interpolations is illustrated in Figure 7.3. We refer to these as polar parallel coordinate plots.

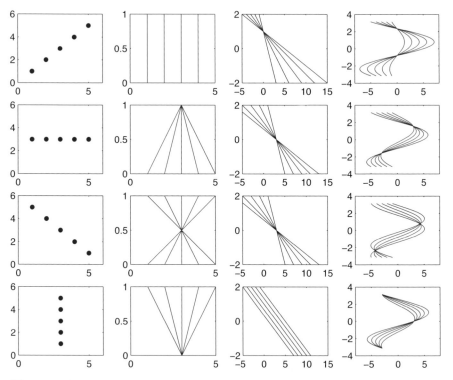

Fig. 7.3. *The data are in column 1. Their parallel coordinate plots using two independent normals are in columns 2 and 3, with the polar view in column 4.*

7.4 A New Family of Smooth Plots

The discussion in Section 7.3 suggests that a more general idea of inter-polation for parallel coordinates is possible and perhaps even desirable. The additional computational overhead is minimal and alternatives may prove to add insight in data exploration. Consider $n(\theta) = (\cos^k(\theta), \sin^k(\theta))$ $0 \le \theta \le \pi/2$. Notice as before $n(0) = (1,0)$ and $n(1) = (0,1)$ forming the orthonormal basis for $\mathcal{R}^2$. Letting $g(\theta) =< \xi_1, n(\theta) >$, notice that

$$g(\theta) = x_1 \cos^k(\theta) + x_2 \sin^k(\theta)$$

$$\Rightarrow g'(\theta) = -k\,x_1\,\cos^{(k-1)}(\theta)\sin(\theta) + k\,x_2 \sin^{(k-1)}(\theta)\cos(\theta)$$

Thus for $k \ge 2$, $g'(0) = 0 = g'(\frac{\pi}{2})$. This means that the interpolation will be locally orthogonal to the coordinate axes. Consider an example adapting this idea for a five-dimensional point. Let $\xi = (x_1, x_2, x_3, x_4, x_5)$ and consider the following

$$g_0(\theta) = x_1 \cos^k(\theta) + x_2 \sin^k(\theta),\ 0 \le \theta \le \pi/2$$

$$g_1(\theta) = x_2 \sin^k(\theta) - x_3 \cos^k(\theta),\ \pi/2 \le \theta \le \pi$$

$$g_2(\theta) = -x_3 \cos^k(\theta) - x_4 \sin^k(\theta),\ \pi \le \theta \le 3\pi/2$$

$$g_3(\theta) = -x_4 \sin^k(\theta) + x_5 \cos^k(\theta),\ 3\pi/2 \le \theta \le \pi.$$

This pattern can obviously be repeated for dimensions $d \ge 5$ simply by repeating the above sequence, i.e., $g_m(\theta) = g_{m+4}(\theta)$ with the appropriate x-components substituted. In general, the $g(\theta)$ functions satisfy the following conditions:

1. Continuity, i.e., $g_m((m+1)\pi/2) = g_{m+1}((m+1)\pi/2)$, $m = 0,1,2,\ldots$

2. Smoothness, i.e., $g'_m((m+1)\pi/2) = g'_{m+1}((m+1)\pi/2) = 0$, $m = 0,1,2,\ldots$

3. Existence of the second derivative, i.e., $g''_m(t)$, $m = 0,1,2,\ldots$

Consider now the following *norm-reducing property*. Suppose $\xi_1 = (x_1, x_2)$ and $\xi_2 = (y_1, y_2)$ are two observations. Let $g^1(\theta) =< \xi^1, n(\theta) >$ and $g^2(\theta) = < \xi^2, n(\theta) >$ and use the $\mathcal{L}_2$ norm.

Theorem: $|g^1(\theta) - g^2(\theta)| \le ||\xi^1 - \xi^2||_2$

Proof: $|g^1(\theta) - g^2(\theta)| = | < (\boldsymbol{\xi}^1 - \boldsymbol{\xi}^2), \boldsymbol{n}(\theta) > | \leq \|\xi^1 - \xi^2\|_2 \|\boldsymbol{n}(\theta)\|_2 \leq \|\xi^1 - \xi^2\|_2$.

The first inequality follows from the Cauchy-Schwartz inequality and the second from the fact that the norm of $n(\theta)$ is less than 1 except at integer multiples of $\pi/2$, where it is exactly 1.

Unfortunately, the inequality does not hold in the opposite direction. Ideally, one would like to have the Cartesian norm preserved in the parallel coordinate domain. However, as with the usual linear interpolations, there is a distance shrinkage between coordinate axes. It should be noted that it is often claimed that because the Andrews plot is a Fourier-like expansion, the Parseval relationship guarantees that $\mathcal{L}_2$ distances are preserved from d-dimensional Cartesian space to Andrews plot space. This result is only valid when there are orthonormal basis sets. The expansion terms for the Andrews plot are not an orthonormal basis.

7.5 Examples

7.5.1 Automobile Data

As an introductory small example, consider data on 74 1979-model-year automobiles; these data are used to illustrate the power of parallel coordinates in visualizing multivariate data and the advantages of the new interpolations when combined with the alternative normal vectors (axis reversing techniques). The dataset has five measured variables (price, miles/gallon, gear ratio, weight, and cubic-inch displacement). For five-dimensional data, there were three distinct pairwise permutations considered in Wegman (1990). They were: Perm_1 = {gear ratio, displacement, price, weight, mileage}; Perm_2 = {displacement, weight, gear ratio, mileage, price}; Perm_3 = {price, gear ratio, mileage, displacement, weight}. These three sets of permutations are shown for the test cases. In the following examples, $k = 3$ has been chosen.

Figure 7.4 shows the traditional parallel coordinate plot where the broken lines represent five-dimensional observations. There are negative correlations between gear ratio and displacement and, similarly, between weight and mileage. This is because it is easy to perceive the heavy crossing of lines with negative slopes between the corresponding pairs of axes. In Figure 7.5, an alternative normal vector, $n_2(t)$, has been used, which rescales the data from -1 to $+1$. Now there is evidence of a linear relationship between price and weight. Using a smooth plot (Figure 7.6) may be more comfortable to the eye and it is easier to follow the curves than the broken lines.

Moreover, the curves cross the axes orthogonally, which gives a clear view of the quantization of the data points. Hence, more information may be obtained from the graphic.

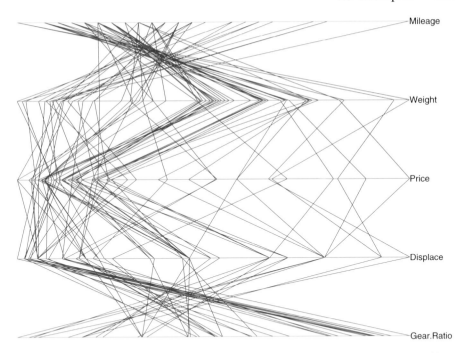

Fig. 7.4. *Parallel coordinate in* 5-*D: Visualizing automobile data using* $n_1(t)$.

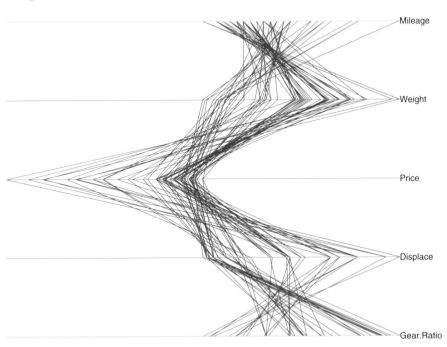

Fig. 7.5. *Parallel coordinate in* 5-*D using* $n_2(t)$ *to flip the price variable.*

Fig. 7.6. *Smooth parallel coordinate in 5-D showing axes quantization.*

Other permutations and alternative normal vectors may be considered to explore further facets of the data. Limited space precludes an extensive exploration. This dataset was already studied in Wegman (1990), and the conclusions are in agreement with what was found then. For example, a heavy car tended to have a large engine providing considerable torque, thus requiring a lower gear ratio. Conversely, a light car tended to have a small engine providing small amounts of torque, thus requiring a higher gear ratio. We believe that smooth parallel coordinates allows us to detect these relationships more easily than the traditional parallel coordinate plot by looking at the axes and between them. It is desirable to have a clear view of the crossings at the axes to reveal quantization and the separation between classes.

7.5.2 Hyperspectral Data: Dealing with Massive Datasets

One of the emerging issues in visual Data Mining is the problem of dealing with massive datasets. The data discussed in this section were developed from hyperspectral imagery taken over the Naval Surface Warfare Center in Dahlgren, Virginia.

The hyperspectral imagery is based on 256 by 256 spatial imagery in 126 spectral bands yielding 8,257,536 pixel values. Not all pixels were useful, so the dataset here uses only 41,748 pixels in each band instead of the 65,536 possible pixels. The exercise for which these data were col-

Fig. 7.7. *Standard parallel coordinate plot of the first 20 principal components of the hyperspectral imagery.*

Fig. 7.8. *Smooth parallel coordinate plot of 20 principal components of the hyperspectral imagery.*

lected was to develop classification algorithms for seven categories of surface objects. These were runway, water, swamp, grass, scrub, pine, and oak. A complete discussion of the analysis process is contained in Wegman and Solka (2005). Dimension reduction was used, taking the first 20 principal components, labeled in the plots as PCA1 through PCA20. The 20 principal components coupled with the 41,748 pixels yield 834,960 numbers, almost enough to legitimately discuss seeing a million. In fact, parallel coordinate displays in CrystalVision, a commercial version of ExploreN, have easily handled 250,000 observations in eight dimensions.

Figure 7.7 represents a standard parallel coordinate plot with linear interpolation between the parallel coordinate axes. Figure 7.8 is based on the same data, but with smooth interpolation between the axes. Both images use desaturated brushing to ameliorate overplotting. It is clear that the feature of the smooth plots that has the interpolants orthogonal to the axes makes low–density regions along the axes substantially more apparent, for example, in the PCA2 through PCA6 axes and again in the PCA12 through PCA16 axes. An interesting added bonus that was not anticipated is that the refreshing time (i.e., the time to redraw the graphic) is four times as fast with the smooth interpolants.

7.6 Detecting Second–Order Structures

In this section, a method is demonstrated that can be effective in detecting more than linear (first–order) structure. The main equation for the method in two dimensions is $g(\theta) = < \xi_i, n(\theta) >$. Another power with the same normal vector can be used to find higher-order crossings in the parallel coordinates. For example, the second order calculation is $g(\theta) = < \xi_i^2, n(\theta) >$. Smoothed plots will even detect data on a circle centered at the origin.

The parallel coordinate plot of a sphere is shown in Figure 7.9 (top left). Its "power of two" visualization is shown in Figure 7.9 (top right). Here, the lines are crossed, which shows that these data come from quadratic structure, and the point of crossing represents the center of the sphere. In Figure 7.9 (bottom left), the polar parallel coordinate plot of the same data is shown and in Figure 7.9 (bottom right), the "power of two" version of the smooth polar parallel coordinate plot. One can easily see that the quantization on these plots reveals more information about the data. Higher-order structures can be obtained by having the power changed appropriate to the cases of interest. Moreover, one can always center by subtracting the mean from each observation, or in general use the z-scores to normalize the data locally or globally as a step in data preparation for the visualization process.

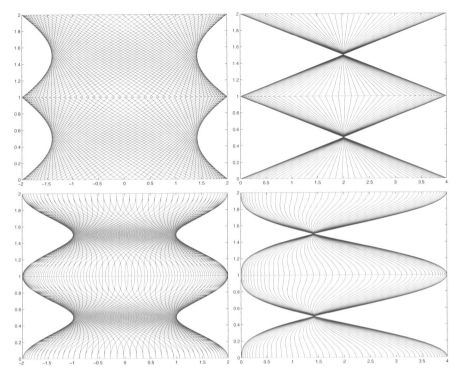

Fig. 7.9. *Standard parallel coordinate plot for a sphere (top left); second-order parallel coordinates for a sphere (top right); smooth parallel coordinates for a sphere (lower left); second-order smooth parallel coordinates for a sphere (lower right).*

7.7 Summary

This chapter has introduced a smooth modified version of the parallel coordinate plot. The modifications are based on a parameter transformation process and its geometric structure. The mathematics behind the new plot has been explained with views that show how patterns may be detected in a dataset. The smooth curves have several significant features, including a norm-reducing property and orthogonal crossings of the axes.

   Although not explicitly mentioned, the analysis of the datasets in the two examples needed a lot of interactions with the software. Actions like reordering and rescaling of axes (cf. Section 4.4.3) or the application of density estimation procedures are necessary steps towards a meaningful and presentable visualization.

8

Networks

Graham Wills

And the marsh is meshed with a million veins.

Sydney Lanier, *The Marshes of Glynn*

8.1 Introduction

Many datasets have special structures that can be represented by graphs of nodes and links, for instance telephone traffic data, trade data, or gene ontology data.

In the mid–1990s, graphs with 100 nodes were considered "large" and ones with 10,000 nodes were considered huge. Little attention had been paid to such-sized graphs, and ones with two more orders of magnitude had not been considered. Thus, most of the existing graph-layout literature concentrates on high-quality layouts of small graphs in various ways and ignores large graphs. The difference between displaying networks with 100 nodes and displaying one with 1,000,000 nodes is not merely quantitative, it is *qualitative*. Layout algorithms suitable for the former are too slow for the latter, requiring new algorithms or modified (often relaxed) versions of existing algorithms to be invented. Iterative layout methods often have stopping conditions defined in terms of a "goodness" measure on the graph, which for very large graphs might lead to expected run times of millennia. Instead an approach which allocates a certain amount of time for a layout algorithm to run is more appropriate.

Even given a good layout, very large graphs require more attention simply to present them on a screen. The density of nodes and edges displayed per inch of screen real estate requires special visual techniques to filter the graphs and focus attention. It is not possible to display an entire unaggregated million-node graph and hope to see useful information. Compounding the problem is that large real-life networks are often weighted graphs and usually have additional data associated with the nodes and edges. A system for investigating and exploring such large, complex data sets needs to be able to display both graph structure and node and edge attributes so that patterns and information hidden in the data can be seen.

This chapter presents a set of techniques for dealing with very large graphs. The emphasis will be on exploration and discovery, not on presentation quality layouts. It will highlight techniques that allow people to interact with very large graphs to

- understand overall structure;
- reveal unusual features;
- compare subgraphs by both structure and node/edge attributes;
- "drill down" from a high level overview to sections of interest.

This chapter will also describe a tool that addresses some of these needs, the NicheWorks tool, which was created by the author while at Bell Laboratories. The images presented in this paper were re-created for this book using the SPSS Visualization system, which incorporates similar graph layout algorithms to those used in NicheWorks. This chapter will further present an overview of recent research on laying out very large graphs and talk about languages for describing graphs and layout algorithms.

8.2 Layout Algorithms

Many papers have been written on ways to lay out graphs. Although not including the most recent work, Di Battista et al. (1994) is still an excellent introductory reference to this literature. The proceedings of the yearly Graph Drawing conferences (e.g., Pach; 2004) are an excellent resource for cutting-edge research and modern methods. The traditional graph-drawing literature concentrates on laying out graphs in two dimensions, but there is also a developing body of work on laying out graphs in three dimensions, which has been created more from the discipline of computer graphics than of mathematics. The results have been rather mixed. The most successful method for visualizing graphs in 3-D, cone trees (Robertson et al.; 1991), depends on a strict hierarchical structure — trees. They are essentially a 3-D generalization of the basic radial layout given below in Section 8.2.1, with clever interaction features for managing larger trees. These trees seem to work well for tens of thousands of nodes, and use of a filtering and pruning interaction mechanism would allow their use for million-node graphs. In contrast, less structured graphs have yet to be displayed successfully in 3-D. It appears that the difficulties of navigation and scene clutter overwhelm 3-D's inherent ability to reduce overlap.

The rest of this section describes three algorithms that can be used to lay out very large graphs, either directly or by applying some modifications. This is not by any means intended to be an overview of the field;

it is instead meant to highlight popular approaches and give an idea of the challenges. Among others, Coleman (1996) gives a list of properties towards which good graph layout algorithms should strive. The list includes notions of clarity, generality and the ability to produce satisfying layouts for a fairly general class of graphs. Speed is also a criterion. However, much of the research into "good layouts" assumes graphs with no data or weights on the nodes or edges. This is unfortunate as, for example, it might matter a lot less if a low-weight node has been occluded than if a high-weight one has. Despite this serious weakness from a statistical viewpoint, the basic layout principles can still be used as guides.

8.2.1 Simple Tree Layout

Di Battista et al. (1994) give several examples of radial layout algorithms of which this one is an example. The tree layout algorithm is designed to work for trees, but also works well for directed acyclic graphs and has proved to be useful for both general directed and undirected graphs. In essence, to use it for a general layout, the "best" tree embedded within the graph is found and that is used for the algorithm. A simple, but effective solution is to use standard algorithms to find a spanning tree within a general graph. For weighted graphs, inverting the weights and finding the minimal spanning tree gives an embedded tree that includes each node's most-heavily weighted edge. Other possibilities, such as creating a complete graph by summing path distances between nodes and using those, are computationally infeasible for very large graphs.

The root node of the tree is positioned at the center and given an angle of 360 degrees. This indicates the angular span of its subtree. The following recursive layout method is then carried out: For each of the leaf nodes of the positioned graph, the angular span available to its sub-tree is divided up using the size of each of its children's sub-trees as weights. Thus, if there was a node with angle 20' and three children with sub-tree sizes 3, 2, and 5, the respective angles allotted to them would be 6, 4, and 10 degrees, respectively. The children are placed on a circle with radius proportional to their distance from the root node and are placed at the midpoint of their

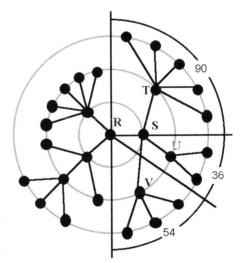

Fig. 8.1. *Radial placement.*

individual angular ranges, with their parent in the center of the overall range (complying with a common criterion for hierarchical layouts mentioned in, for example, Coleman (1996)). An example is shown in Figure 8.1. The root node (R) is drawn at the center, with its children on a circle centered at R of radius l. R has a sub-tree of size 20 and its child S has a sub-tree of size 10, so S acquires an angular span of $360 \times 10/20 = 180°$. Its child T with sub-tree of size 5 gets a span of $180 \times 5/10 = 90°$, U gets $180 \times 2/10 = 36°$, and V gets $180 \times 3/10 = 54°$.

This process continues until all the nodes have been positioned. The order of placing sub-nodes around the circle is the final element requiring definition. The approach used here is to place the strongest weighted edges from a node to a child node in the center of the span, with the weaker ones to the edges. Thus, the child with the strongest edge to its parent will be in the center of the range. This works in line with a general principle that the shortest edges should indicate the strongest weights — in other words, nodes that are linked by a highly weighted edge should be closer to each other than ones linked by a lesser weighted edge.

In practice, a slight modification to this algorithm, which does not use all the available angle to place children, improves the layout. The slight loss in available space is more than offset by the improvement in visible separation between subtrees. Figure 8.2 shows a very small tree layout under this system.

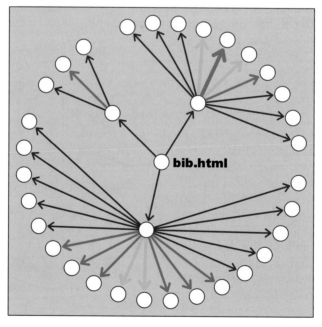

Fig. 8.2. *A small radial tree layout with additional separation between sub-trees.*

8.2.2 Force Layout Methods

Force Layout Methods form a large and general class of layout algorithms. They are intuitive and a naïve version of the algorithm is simple to program. Although the details vary widely, the basic approach is to form a measure of attraction or repulsion between all pairs of nodes in a graph, based on the graph edges, and then solve the resulting physical model, either by solving a (typically) matrix equation, or by starting with an initial configuration and simulating the physical process. By choosing the way to express edge weights as forces, two important sub-methods emerge:

- **The Spring Model**
  This model uses the edge weights to define a virtual spring connecting the nodes. These springs have a force acting on them that obeys Hooke's law for springs, so each "spring" will have a desired length and will exert a force between its terminating nodes to bring them to that distance apart. This model is solved by simulation.

- **Multi-dimensional Scaling**
  In this case, the forces are described so that the statistical multidimensional scaling (MDS) algorithm applies. Essentially this requires a symmetric matrix to be created describing the preferred distance between each pair of nodes and node locations are found to minimize the squared error between the resulting distances and the preferred distances.

Classical MDS is a non-starter for large graphs, because it requires matrix solutions with matrices of size N^2, where N is the number of nodes in the graph. Solving these matrices when $N = 1,000,000$ is infeasible. Techniques have been proposed for generalizing MDS and not requiring a matrix solution, but using iterative solutions; these essentially boil down to solving a general force based model, as discussed at the start of this section.

One other disadvantage of these models is that to get good results, it is often necessary to include a force by which nodes repel all other nodes to which they are not linked by an edge. Rather than add all the extra forces in directly, which would make the number of forces N^2, only those nodes sufficiently close to each target node need be considered. This reduces the number of forces to be considered to $N \log(N)$ for a wide number of common ways for dividing up a space into local sections. For example, Nievergelt and Hirichs (1993, Section 23.3) use a quad-tree algorithm that is $O(\log N)$ for all three operations of adding, deleting and calculating close neighbors, leading to the overall $N \log(N)$ performance.

8.2.3 Individual Node Movement Algorithms

This algorithm is designed to improve a random layout rapidly; it is especially useful for random-grid layouts. Again, forces are defined between
all pairs of nodes and the sum of all these forces for a given configuration is termed the potential. These algorithms pick a node and then see if
moving just that node will decrease the overall potential. If it does so, the
node is moved. A variation, based on simulated annealing, allows small
potential-increasing movements so as to break out of local minima. Over
time, the maximum allowed size of these small potential increases is decreased, simulating a lowering of temperature in a physical system as the
configuration "freezes" into position.

One version of this algorithm randomly picks a pair of nodes and cal-
culates the difference in potential if the nodes were swapped. If the swap
decreases the potential, then the nodes' positions are swapped. This has
been used in NicheWorks. A similar algorithm can be found in Davidson
and Harel (1996). They use an annealing approach to decide whether to
move a node to a new randomly chosen position near the starting one.

One problem with this method is in adapting the recommended number of iterations and the cooling schedule. Davidson and Harel (1996)
suggest $30N$ iterations, which is impractical when $N = 1,000,000$.

Calculating the effect on the potential of a swap is linearly dependent
on the number of edges involving either node. Sparse graphs are more
common for large networks than near-complete graphs, with the average
degree of the nodes remaining nearly constant as more nodes are added
to the network (this reflects the author's experience with a range of different data sets: mainly telephony, software call graphs and modification
histories, and text document associations via n-gram analysis). Thus the
potential calculation is typically very fast and can be performed many
times. Because nodes can be moved very long distances with one swap,
this method is a powerful way of improving random layouts rapidly. However, a bad initial placement can lead to very poor long-term performance
and becoming trapped in local minima is still an issue, even with annealing.

8.3 Interactivity

Ben Shneiderman's information seeking mantra (Shneiderman; 1996) of
"overview, filter and zoom, details on demand" is highly applicable to the
exploration of very large graphs. It is simply not possible with current
technology to look at a million nodes with all their links on a display
and see anything useful. In my research, I have used a graph visualization system within a linked views environment. Such an environment is

described in detail in Wills (1997) and Eick and Wills (1993) and implemented at least partially in systems such as Velleman (1997), Wills, Unwin, Haslett and Craig (1990), Swayne, Cook and Buja (1991), Tierney (1991), and Unwin, Hawkins, Hofmann and Siegl (1996). Under this paradigm, each view of the data is required to represent both the data themselves and a state vector that is attached to the data. This state represents the "interest" in each data item and forms a way both to focus attention and show/hide sections of views. In the NicheWorks and EDV implementations Wills (1997) the possible states are

- **Deleted**
  Treat the data point as if it were not present.

- **Normal**
  Show the data.

- **Highlighted**
  Show the data so it will stand out against *normal* data.

- **Focused**
  Show as much detail as possible of the data.

Furthermore, the user should be allowed to modify the state vector by interacting with the data views. For example, selecting a specific bar from a bar chart view and highlighting it will change the data state vector for items represented by that bar, causing other views of the data immediately to update their representation. This paradigm allows the user to interact with a graph view of the data in conjunction with views of any node or edge attributes. The following operations are then made available:

1. Tools to colour and represent nodes and edges based on their attributes.
2. Selective labeling of nodes under user control; also painting of nodes with labels.
3. Mapping of attributes to shapes and labels.
4. Interactive data querying via the mouse.
5. Ability to show or hide parts of a graph via manipulation of views of node/edge attributes.

These capabilities are shared by all tools in a linked views environment. There are also methods specific to graph analysis that are incorporated into the NicheWorks view. These include:

6. Selection propagation from nodes to edges and vice versa.
7. Selection propagation within a graph by following edges (one step or connected component).
8. Interactive Pan/Zoom and Rotate facility.

Operations (1) through (3) enhance the basic views with additional information. For example, the shape of a node may encode one attribute while colour encodes another, allowing a multivariate decoration of the basic graph structure. For a weighted graph, it is almost mandatory to represent the weights for the drawn edges in some visual way. Colour or line thickness can be used, but with modern drawing environments more subtle effects, such as transparency (with anti-aliasing) provide an effective alternative. (4) provides details on demand, and is especially useful for displaying text information which cannot be easily encoded, such as a person's name or a web URL.

(5) Allows logical zooming to be performed. It is probably the most important general technique. It allows the user to display, for example, just the relationships between webpages that contain text rather than images; to hide all the low-importance nodes; to see only those edges that exceed a certain threshold; or to display selected categories of nodes. If attributes are available for both nodes and edges, the graph can be used to provide a way of propagating selections from node attributes to edge attributes and vice versa, as indicated by (6) and (7). In analysing web traffic, first select all nodes that represent a "decision to purchase" page, propagate that selection to the edges that are directed into that node and also to the nodes at the originating end of those edges. Then focus attention on those webpages that lead to purchase decisions and the links that were used to get to that decision.

Even when no extrinsic attributes have been specified for nodes and edges, many graph-theoretic attributes can be calculated based purely on the layout. The number of edges originating at a node, O, provides one, as does the number of edges, T, terminating at a node. The expression $(O - T)/(O + T)$ gives a measure of whether the node is a source or a sink, on a scale of $(-1, 1)$. For edges, attributes can be calculated such as whether an edge separates any nodes in a graph or the number of minimal paths between nodes that use that edge. These measures of importance for both nodes and links can then be used for highlighting, encoding or for reducing the number of nodes to display. For example, hiding all nodes with only one link at least halves the size of a tree.

8.3.1 Speed Considerations

For large graphs, it is very important that the zoom/pan and rotate interactions are sufficiently fast and also that visual feedback is provided as they are being drawn. Because drawing $1,000,000$ nodes is never going to be a rapid task, how can these two criteria be reconciled?

Research into human-computer systems has shown that if an operation is performed sufficiently fast (under 200 ms for the average person), then it is perceived as being instantaneously connected to the initiating actions. This has important implications for graph drawing programs.

The results of any mouse action, such as zooming in or out, panning around or rotating the graph, should be continuously re-displayed within this time limit. Failure to achieve this means that the user will have difficulty manipulating the view. Of course, redrawing a million nodes and edges of varying shapes and colours within 200 ms is impossible on most machines, and by the time it is possible, the target will be to visualize a billion nodes and edges. Some partial-drawing solution is needed.

A naïve approach would be to start drawing the graph at time t_0 and continuously check the time until time $t_0 + 200$ ms is reached and then stop drawing. Unfortunately, this has a severe failing; for a very large, dense graph, which has a lot of overplotting, the partially drawn graph is guaranteed to look very different from a completed drawing. This is because the partial drawing shows only those nodes and edges that lie *below* the ones that are usually drawn later and so appear on top.

In NicheWorks, a statistical technique was employed to solve that problem. You want to be able to predict the number of nodes that can be drawn in 200 ms. Then start drawing that far from the end of the list of items to draw and you guarantee both to finish on time and to have drawn all the nodes that normally appear on top in the full rendering. Of course, the amount of drawing that a computer can do in a fixed amount of time varies quite wildly on a multi-tasking system, so a robust statistical prediction needs to be made based on the current history of drawing. Essentially, at any point there is a history of number of nodes drawn (N_t) and the time to draw those nodes (T_t). There will be a functional relationship between them, and knowledge of the drawing algorithm leads to trying a function approximately linear in the relationship between T_t and N_t.

The details of the prediction algorithm are quite complex, as it needs to be robust, especially to the sudden spikes associated with processor sharing, and yet smooth, as nodes and edges should not flicker during zooming and panning. It must also be adaptive, because zooming and panning mean that different numbers of nodes and edges have to be drawn and the system takes longer to draw graphic objects of differing sizes. After a large amount of experimentation, the technique that worked best was to use weighted regression where weights decreased exponentially for records occurring further in the past. Three such regressions were calculated, each leaving out one record of the last three and the median value used. Empirical results on UNIX X-Windows systems and Windows/DOS boxes indicate that the prediction method works well, rarely overshooting more than 50% of the time and without too much flicker.

8.3.2 Interaction and Layout

The state vector can be useful when laying out large graphs. If a node's state is set to deleted, then it plays no part in the layout process, nor do any edges involving it. Thus, the deletion mechanism can be used to look carefully at subsets, trying layouts only for them, or using partial layouts to speed up positioning a very large graph.

Another use of the state vector is to allow the user to *fix* the selected nodes. These nodes are not permitted to be moved by any of the subsequent layout algorithms. This allows the user to layout one subset of nodes, then fix their positions and use, for example, the descent method to move the other nodes into the best positions relative to the fixed nodes.

A final consequence seems trivial, but is very useful in practice. By showing only the most important nodes and/or strongest edges, the user can watch the layout algorithm perform without taking too much time away from the algorithm to display the data. This helps the user understand the algorithm's operation more clearly and thus aids the layout process.

8.4 NicheWorks

NicheWorks is a visualization tool written at Bell Laboratories to investigate very large graphs. Typical analyses performed using NicheWorks have between $20,000$ and $1,000,000$ nodes. On current mid-range workstations, a network of around $50,000$ nodes and edges can be visualized and manipulated in real time with ease. Increased size decreases interactive performance linearly. NicheWorks allows the user to examine a variety of node and edge attributes in conjunction with their connectivity information. Categorical, textual and continuous attributes can be explored with a variety of one-way, two-way and multi-dimensional views.

NicheWorks was originally designed to examine large telecommunications networks, and has been applied extensively to understanding calling patterns among telephone customers both for customer understanding and fraud detection. In this domain, information about the customer (type of service, geographical location, etc.) and the calls they make (date, time of day, duration) needs to be understood in the context of who calls whom. Typical questions that NicheWorks was designed to answer deal with clustering users based on their calling patterns and collected statistical attributes; detecting and characterizing atypical calling patterns; understanding how an interesting subset's calling patterns differ from those of the whole; and identifying individuals who have been classified into a given category by a purely data-driven algorithm, but who do not appear to fit that category based on their calling patterns. NicheWorks has been described in detail in Wills (1997).

8.5 Example: International Calling Fraud

Discovering fraudulent calls made to overseas locations is very important for telephone companies. Not only do they lose revenue if they cannot collect money for a call, but they also have to pay overseas telephone companies to use their equipment, resulting in an actual loss of money. But telephone fraud is very resilient to automatic or Data Mining techniques for detection. Fraudsters adapt very rapidly to new algorithms and share their knowledge on preventive systems. For this reason, a visual approach to fraud detection is very effective; it exploits users' ability to interpret and understand new patterns in calls, allowing them to process more data and identify new fraudulent methods. The data set in this section originally consisted of more than 20 million international telephone calls over a weekend in 1994. This has been segmented by geographical origin, and here a moderate $80,000$ calls are investigated, involving slightly less than $35,000$ callers. Figure 8.3 shows the results of the default NicheWorks al-

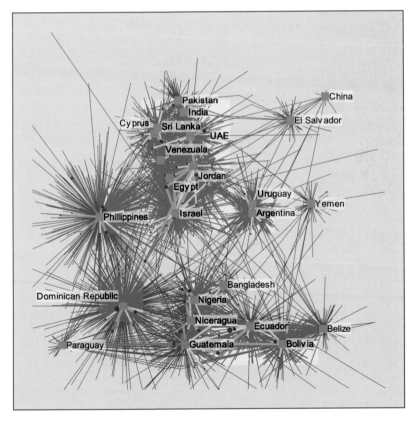

Fig. 8.3. *Overview of calling patterns.*

gorithm. This algorithm places nodes randomly on a hex grid and then swaps them around using simulated annealing until few further changes are made. Then the algorithm performs steepest descent minimizing the "badness" of the layout (its potential) as a function of the node positions. A final step moves all nodes slightly apart from their neighbors where that is possible. The potential function used and details of the algorithm are described in Wills (1997).

The most noticeable feature of Figure 8.3 is the number of callers clustered around each country; these are people who called exactly one country, with the distance to the country indicating for how long they were on the phone to it. In Figure 8.4, the links corresponding to long call times have been selected and that selection propagated to the nodes at the ends of those links. The rest of the graph has been hidden and the resulting

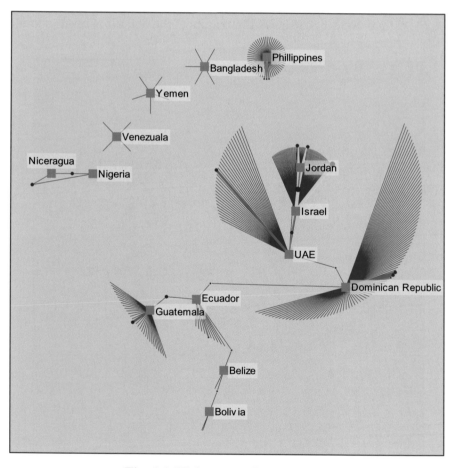

Fig. 8.4. *High users' calling patterns.*

graph laid out using the radial tree layout on the maximal embedded tree as described in Section 8.2. Hiding of links has caused the graph to resolve into six components; these are laid out individually, and the resulting layouts composed together.

The analyst noted an interesting couple of callers who called both Israel and Jordan a lot. By using the mouse to focus on the callers, and viewing the results in a linked view, it appeared that each of the callers made more than 120 calls to Israel and more than 80 to Jordan. A fisheye transform was applied (a distortion view that magnifies around the area of interest and shrinks elsewhere), producing Figure 8.5. The two nodes of interest are shown in the diagram linked with large yellow links to Israel, and with medium-strength cyan links to Jordan. When mousing over the nodes for information, it was apparent that not only were their calling

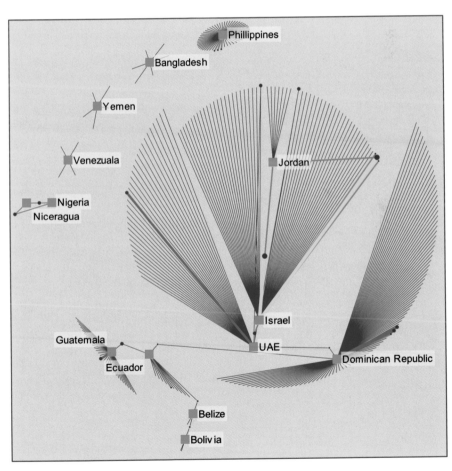

Fig. 8.5. *Possible fraud pattern involving Israel and Jordan.*

patterns very similar, but they also had very similar numbers (which have been scrambled here). It is almost certain that these callers are part of the same operation. At the time of the study, it was impossible to call directly between Israel and Jordan. Some companies would set up a phone account in a rented apartment in the US and charge Israeli and Jordanian business people for third-partying the call through to the other country. When their bills came in at the US end, they would simply ignore them and leave to set up a new location. The distribution of durations of calls made by these numbers is consistent with the investigator's expectations for this type of fraud.

Looking at these callers in detail, the analyst noticed that they also called the United Arab Emirates (UAE) occasionally. Following this line of inquiry, the analyst showed all calls and callers involving any of these three countries and produced Figure 8.6.

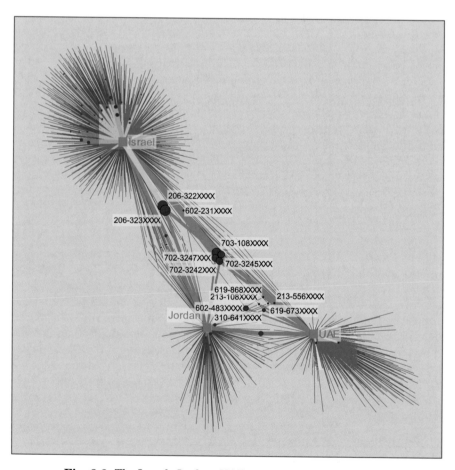

Fig. 8.6. *The Israel–Jordan–UAE generated subset: overview.*

By labeling interesting numbers, the analyst identified one number that called Israel for a long time and, importantly, a group of other telephone numbers having similar patterns of calling to the original pair and having numbers similar to each other (the numbers beginning with 702-324). This set of numbers had much lower call volume, partly because there were more phone numbers involved, and so were less easy to detect. Because standard methods would not have noticed that the numbers were similar, they would not have been able to identify this type of fraud.

Using the NicheWorks tool, the analyst was able to explore a medium-sized network of more than 40,000 calls, using different criteria to selectively show different aspects of the data and using different positioning to highlight information.

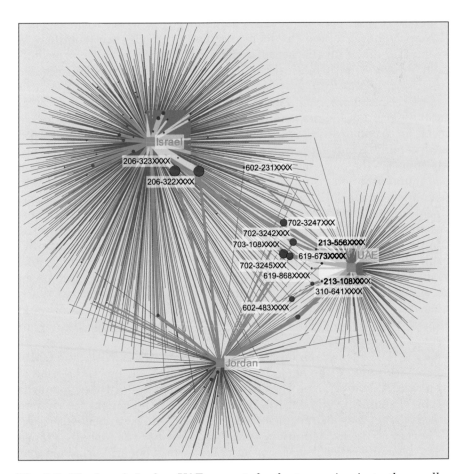

Fig. 8.7. *The Israel–Jordan–UAE generated subset: zooming in to those callers calling more than one country.*

8.6 Languages for Description and Layouts

One of the challenges in working with graphs is diversity. There are many algorithms that can be used to lay out graphs, and these algorithms can be combined. Given a layout and data, graphs can be displayed in many different ways. Even the definition of a graph — nodes, edges, attributes — can be annotated with other information on a priori hierarchies and graph attributes that complicate the basic simple structure. In this section some recent approaches to coping with this diversity by using descriptive languages are discussed.

8.6.1 Defining a Graph

At its most basic, a graph consists of a list of edges, where each edge has a "from" identifier and a "to" identifier. The set of all such identifiers is assumed to make up the nodes of the graph. Whereas this can be sufficient for some mathematical applications, real-life applications require more work.

One recent approach is GraphXML (Herman and Marshal; 2000), a graph description language in XML that can be used as an interchange format for graph drawing and visualization packages. The generality and features of XML make it possible to define an interchange format that not only supports the pure, mathematical description of a graph, but also the needs of information visualization applications that use graph-based data structures. GraphXML allows hierarchies and dynamic graph features to be described, as well as allowing for some number of display options. Care has also been taken to allow it to be extensible. From the viewpoint of large graphs, however, it has the disadvantage of requiring that the graph data be embedded in the XML, which for $1,000,000$ node graphs means a lot of unnecessary text parsing and translation. A mechanism that allows the structure to be defined with a reference to the data (such as used in VizML, Section 8.6.2), rather than conflating the two would mitigate this problem and would not be hard to integrate into the overall design.

Other XMLs for graph description include XGMML, GXL, and GML. They share some of the same features as GraphXML but miss some of GraphXML's innovations. They all share the same drawback for large graphs of requiring the data to be embedded in the graph description. As a side note, they also require style information to be embedded as part of the core structure definition, which makes it hard to globally change styles or modify sets of items. For large datasets, a general principle that should be considered is to separate the data elements, the structure elements and the style elements as much as possible so that changes to one do not require computation over all of them. Changing the font for node titles in a 1,000,000 node graph should not require a change in the XML for each of those nodes. Similarly, an update to a database entry should not require regeneration of a multi-megabyte XML document.

8.6.2 Graph Specification via VizML

The specification of graphs is part of a larger problem, the specification of visualizations in general. Many attempts have been made to create taxonomies of visualizations, but these have tended to be at the rather superficial level of categorizing by dimension and by the appearance of the main symbols (bars, lines, points, etc.). SPSS has been developing an XML variant for general visualization definition. The purpose is to provide a way of defining the essence of a visualization, not ways to create "barcharts", "scatterplots", or other predefined, canned views. As one of the goals is to be able to reproduce all common visualizations, this XML can be used to specify graphs.

VizML provides a means of defining a visualization (a.k.a. graph, plot, view, or chart) by specifying the components that make up a chart, such as frames, points, axes, and the like. At the lowest level a chart can be thought of as a collection of lines, symbols, and text and graphical attributes for these objects. This level is lower than VizML. At the highest level is a specification of a chart by type. This specification is very good for initial chart creation (I want a scatterplot of height vs. weight) but is very limiting in that every desirable chart must be pre-conceived and planned for, and all options for that chart must also be thought of in advance. If no one ever thought of adding a smooth line to a histogram, it simply cannot be done.

VizML lives between these two extremes. It defines high-level items, but not complete charts. These items correspond to the "statistical objects" as described in Chapter 4. It also allows these items to be composed in any way that makes any sort of sense. As an example, any chart may be rendered in polar coordinates, which allows a histogram to become a variant of a Rose diagram (as invented by the famous nurse and early applied statistician, Florence Nightingale). If applied to a scatterplot with a loess smoother, it transforms it into a polar scatterplot with a circular smooth. This is one of the core ideas behind VizML providing a set of orthogonal graph concepts (elements, coordinates, aesthetics, transforms, statistics, interactions, styles) that can be composed as flexibly as possible to provide an extremely wide variety of charts. Node and edge displays fit into this framework as two elements (points and links) with a graph layout statistic on them to lay out the graph. VizML is based on Lee Wilkinson's *The Grammar of Graphics* (Wilkinson; 2005), which provides more details on VizML. In this chapter, the colour figures were all produced by VizML. For example, here is an XML fragment from the specification used to produce Figure 8.5:

```
<graph>
 <coordinates>
  <dimension lowerMargin="5%" upperMargin="5%"/>
  <dimension lowerMargin="5%" upperMargin="5%"/>
  <fishEye magnification="5" x="0.666" y="0.63"/>
 </coordinates>
 <link>
  <graphLayoutStatistic from="from" to="to" weight="weight">
   <initial type="polarTree"/>
  </graphLayoutStatistic>
  <color high="yellow" low="blue" variable="weight"/>
  <size high="3px" low="0.5px" variable="weight"/>
 </link>
  ...
</graph>
```

It contains a simple 2-D coordinate transformation, with a fisheye transformation chained after it magnifying at a location specified interactively by the user. A link element is defined with a single layout method — just do an initial polar tree layout, then stop. The links are coloured and sized by the link's weight, which is derived from the total call time. Other details of the XML have been omitted for clarity. To produce Figure 8.6, essentially the same specification was made, except that the graphLayoutStatistic has been changed to the following:

```
<graphLayoutStatistic from="from" to="to" weight="weight">
 <initial type="grid"/>
 <swap ignoreLinkCrossings="true" maxTime="60"/>
 <move ignoreLinkCrossings="true" maxTime="90"/>
</graphLayoutStatistic>
```

For this layout, you start with a grid, then swap nodes for up to one minute, then move nodes to better locations for up to a minute and a half. For both these update algorithms, links that cross are not considered in the calculations of better (i.e., when the potential of a layout is calculated, link crossings do not increase that potential).

8.7 Summary

Million-node graphs arise far more often than might be expected. Any large group of individuals or items that interact or communicate in any way will generate networks, and so there are many examples that are important to study. For this, exploratory and presentation tools are needed.

The first issue that must be dealt with is the specification of the graph. What are the nodes? The links? What data are attached to each? How should it be drawn? What attributes dictate the shape of the nodes, their colour, their labels, and so on? Are all the links equal, or are some to be shown stronger, thicker or coloured/dashed in some way? Most existing graph specification languages, by tying together edge variables, positions, appearance, style and meta-information, make it very cumbersome to represent million point nodes. Instead of one conflated specification, this paper suggests that a specification language such as VizML, which separates out the data and styles from structure and information, is superior. It allows the node and link tables to exist in a database or in any other external location, and so allows easy manipulation of the structure.

With the specification defined, layout algorithms must be used to display the graph. Most literature in the field is concerned with graphs of up to a few hundreds of nodes and only a few commercial packages can handle low thousands of nodes. A different class of algorithm is needed which does not aim for perfection, but for producing an informative display within the limited amount of time available. The bound is time, not quality.

With a million nodes, both an overview and the ability to focus and filter on details are necessary. Linking nodes and links to statistical plots of their attributes allows graph structure and data structure to be co-explored, seeing one in the light of the other. Additional selection operations that allow the user to expand selections by selecting nodes and edges connected to ones discovered to be interesting allows questions like "who are all the people who phone people with characteristic X?" to be answered. Once an interesting subgraph is found, allowing layout algorithms to be run interactively on just that subgraph completes the circle by allowing high-quality layouts to be performed on small graphs, so you can take advantage of the large body of existing work on layouts for small graphs, while living in a world of million-node graphs.

9

Trees

Simon Urbanek

Millionaire models are rare enough, but model millionaires are rarer still.

Oscar Wilde, *The Model Millionaire*

9.1 Introduction

Trees have been used in many scientific fields for decision making, classification, or prediction for quite some time. They are usually easy to interpret, even for people without statistical knowledge, because they are so easy to visualize. This makes their application appealing for a wide variety of tasks.

In statistics, tree models are most widely used for classification and regression (Breiman et al.; 1984). Besides their interpretability, they are also very flexible in application, because they are non-parametric and require no assumptions about the distributions of the covariates. All kinds of covariates, categorical, ordinal, or continuous, can be mixed together in the model and they handle missing values well. These properties have made trees popular, especially for non-linear regression and classification problems.

Recent fast-paced developments in computational power have caused the research in this field to go in separate, different directions. The first consequence is the increase in the volume of available data. Automatic data acquisition and the large databases that are commonly in use nowadays offer overwhelming amounts of data waiting to be analysed. This leads to questions as to whether trees are suitable for handling such large amounts of data, as to how they behave when many different covariates are present, and as to how they scale up with increasing numbers of cases. All these issues are naturally important for visualizing trees. If models cannot be calculated quickly, graphical displays cannot be drawn quickly, especially the displays that are based on results from groups of trees.

Unlike many other models, there is a variety of different algorithms and methods for growing trees because of their complex structure. Nothing like "the optimal" tree exists, and it is possible to grow trees of almost any size. The first section discusses how to handle large trees and how tree size is related to the size of the dataset. The bigger the tree, the more difficult it is to visualize and interpret.

There are various ways of visualizing a tree. Visualization is important in order to interpret models and to assess their quality. The rich structure of a tree cannot just be expressed numerically; it gains from graphical displays. Increasing dataset size means that standard plots need to be re-evaluated, their limits determined and solutions for problems induced by the increase in size found. Chapter 3 illustrates these problems and gives possible solutions. All the tree displays in this chapter have been generated using the interactive statistical software for visualization and analysis of trees, KLIMT. KLIMT provides both a broad range of options for displaying and exploring individual trees and a variety of innovative plots for understanding and investigating collections of tree models.[1]

Current computational tools allow the generation of large numbers of trees for a single dataset. One idea is to combine the information contained in all the trees in order to improve prediction accuracy. The loss of interpretability of the combined predictors can be partially offset by using methods for analysis and visualization of the differences and similarities between the models.

The methods for large datasets, large trees, and large numbers of trees outlined in this chapter are summarized in the last section together with their benefits and their limitations.

9.2 Growing Trees for Large Datasets

There are several different methods for tree construction. The well-known ID3/CART family of greedy algorithms is mostly used for initial growing, and then various pruning algorithms are applied.

The Bayesian approach (Chipman et al.; 1998) offers an alternative to CART. CART is based on local optimization, which may be inappropriate for a given task. Bayesian trees are not bound by this restriction. Unfortunately, Bayesian approaches are computationally very demanding. The construction of a Bayesian tree model for a dataset of 600 cases takes several minutes. This rules out its application for large datasets. In the next section, the performance of the greedy algorithm is analysed as to how it is affected by the number of observations and by the number of variables.

---

[1] Tools like SpaceTree (Grosjean et al.; 2002) are useful for displaying large hierarchical datasets but are less helpful for decision tree models.

9.2.1 Scalability of the CART Growing Algorithm

Let $X_1, \ldots, X_p$ denote the predictor variables and N the number of cases. Consider the computational steps necessary for growing a single tree.

The idea of the greedy algorithm is to consider all possible splits for each variable, calculate the decrease of impurity for each split, and choose the one with maximal decrease of impurity. The measure of impurity used here is entropy, which up to a constant factor corresponds to the deviance.

Computational complexity depends on the kind of variable used in a split. Take continuous and ordinal variables first. These variables can be ordered by value. Each split in a node t partitions the cases $\{x_{i1}, \ldots, x_{iN}\}$ falling into the node into two groups according to $x_{ik} \leq s_t$ and $x_{ik} > s_t$, where x_{ik} denotes the k-th case of the variable X_i and s_t the split cutpoint for the node t.

Therefore, the naïve approach would involve calculating the deviance gain for each of up to $N - 1$ splits. For regression trees the deviance is defined as $\sum_{k=1}^{N} (y_k - \hat{y}_k)^2$. Regression trees with a constant model in terminal nodes (or leaves) are characterized by $\hat{y}_k = \bar{y}_t$, where t is the terminal node the case k falls into and $\bar{y}_t$ is the mean of response values of the cases falling into that node. This leads to a computational cost of

$$C_t = C_{\text{mean}} + C_{\text{deviance}} \approx O(2N)$$

for each participating node. This needs to be calculated only once for the parent node, but once for each possible split into two child nodes. Therefore, the global computational cost of finding the best split for one variable is:

$$C_v = C_{\text{parent}} + \sum_{s=1}^{N-1} (C_{t_l s} + C_{t_r s}) \approx O(N^2),$$

which scales badly for large datasets.

Fortunately, it is possible to perform several optimizations. If the search is done in the order of the values of the variable split, then it is possible to update both predicted values and the deviances of the children based on the values of the previous step. This reduces the computational costs to $O(N)$.

Furthermore, if the data are organized in an intelligent way, then it is possible to reduce the computational costs of a split on one variable to $O(L_i)$, where L_i denotes the number of discrete values taken by X_i. Because splits are always placed between two distinct values of X_i, it is sufficient to consider splits between the values, and given the relative counts it is possible to use an updating formula that is based on the values L_i instead of on the cases.

For classification trees the deviance of a node t is defined as

$$D_t = -2 \sum_{\text{classes } c} n_{tc} \log \frac{n_{tc}}{n_t}.$$

The values n_t and n_{tc} can be obtained by induction. This leads to computational costs of $O(CN)$, where C denotes the number of categories in the response variable. A reduction to levels instead of cases is also possible, but the information that needs to be cached includes the changes of n_{tc} for each case and is therefore more complex than in the regression case.

So far, splits on ordinal and continuous variables have been considered. It remains to check the computational costs of splits on categorical variables. Evaluating all splits means calculating $2^{C_i} - 2$ possible alternatives, where C_i denotes the number of categories for the split variable X_i, because each category can go either to the left or the right branch and cases where one of the branches is empty are not of interest.[2]

Given pre-cached information on each category c_i it is possible to use induction update formulas similar to those used for ordered variables. The generation of such a cache pool has complexity $O(N)$, which leads to a global computation complexity of $O(N + 2^{C_i})$.

In many applications, N is the dominating factor: complexity depends on the number of cases linearly. Nevertheless, 2^{C_i} can be large because it grows exponentially with the number of categories C_i. Therefore, where possible, categorical variables with many categories should be converted into ordinal ones or split into separate categorical variables with fewer categories.

So far, the split of a single node with N cases has been considered. A tree is generated by recursive splitting of the nodes. The number of cases across a level of the tree (i.e., nodes of the same depth) is constant until terminal nodes are reached, then it is reduced by the number of cases in the terminal nodes. This implies that for a full tree the number of cases across each level is N, and this is also the worst case considering computational costs given the height h of the tree. The total cost of growing a full binary tree of height h is approximately $O(hpN)$, assuming that for each node and variable the cost of finding the best split for one variable and one node with N_t cases is $O(N_t)$, that p variables are to be considered each time, and that the sum of all N_i across a level is N with a total of h levels.

This means that the entire growing process scales linearly with the factors N, p, and h. For subtrees of the full tree, the computational costs decrease correspondingly. N and p are fixed for a given dataset, therefore the complexity depends only on the number of cells in the dataset and on the height of the tree.

The height h is not directly defined by the data but depends on the stopping rule used for generating the tree and on the amount of infor-

---

[2] Some programs such as Quinlan's C4.5 and C5.0 split nominal variables to C_i branches where each branch corresponds to one category. This greatly reduces the costs of such splits, but they are not often desirable because groups of categories are not allowed.

mation contained in the data. Trees for classification problems tend to be smaller than regression trees, because of the fixed set of possible predicted values in the classification. If the tree manages to separate the classes fairly well, i.e., each terminal node corresponds to one class, then h will be of the order of $\log C$ in the optimal case of a full tree (a full tree with k-way splits of height h has k^h terminal nodes and those correspond to the C classes), or at worst $C - 1$ for an unbalanced tree, i.e., a tree that splits off only one terminal node at each level. Given a large dataset, $p \cdot N$ is by far the most dominant factor.

The conclusion of this analysis is that the greedy algorithm scales approximately linearly with the size of the dataset and so is suitable for modelling large datasets. The number of variables and the number of cases both have the same weight in the computational complexity.

Before continuing with the complexity of pruning methods, consider the evaluation of tree models. A new case is evaluated in a tree model by passing the case down the tree according to the splitting rules along the way. At each node only one variable has to be evaluated. This means that the result is reached after at most h steps, which is very fast for most trees.

It is possible to grow large trees, but arbitrarily large trees are not necessarily useful, and pruning methods are used to prevent overfitting. Good pruning gets rid of uninformative branches and cuts back to the essentials of the tree. The effects of pruning can be seen in outline by overlaying the trees before and after pruning. Of course, the pruned tree can be visualized in detail more easily than the unpruned tree.

9.2.2 Scalability of Pruning Methods

As long as any two cases with equal predictor vectors belong to the same class, it is possible to construct a perfect tree — that is, a tree that classifies every case of the training set correctly simply by growing the tree until all terminal nodes are homogeneous. This tree will be very large because it consists of up to as many terminal nodes as observations. The practical value of the tree will be low, because the misclassification rate for test samples will be high.[3] It is necessary to find a balance between tree size and general prediction accuracy.

One way is to use stopping rules. They define conditions determining when the growing process should be stopped, usually by specifying thresholds for the number of cases in a node and for the impurity decrease of a split. Stopping rules are very easy to implement and computationally cheap because the values are calculated during the tree construction process. Their main disadvantage is that they do not necessarily produce

---

[3] Unless the underlying problem is not noisy, but then the tree is likely to be small and pruning is not needed.

good trees. Sometimes, a tree has one weak split followed by a strong one, because the group of interest can only be separated by using the two splits in combination. A stopping rule could prevent this, if the first split falls below the specified threshold.

Consequently, the reverse method is used — trees are grown to their full size and pruned by removing branches that are weak. There are many pruning methods, and here two will be described, representing both the classical approach and the statistical testing approach: cost-complexity pruning and a statistical testing procedure.

Examining all possible combinations of pruned branches is computationally not feasible, so both methods use the idea of the weakest link. All lowest-level inner nodes are examined, and the one with the smallest contribution is removed. This is repeated recursively until the root node is reached. From the sequence of trees obtained, an optimal one is chosen.

The main differences in the pruning techniques lie in the choice of the best tree and in the method of finding the weakest link. Cost-complexity pruning uses a measure

$$R_\alpha(T) = R(T) + \alpha |T^{term}|$$

where $R(T)$ denotes the estimate of the true misclassification rate of T and $|T^{term}|$ the number of terminal nodes of T. The weakest link is determined by the node minimizing the function

$$g(t) = \frac{R(t) - R(T_t)}{|T_t| - 1}$$

among all lowest-level inner nodes. The branch at this node is pruned and the procedure is repeated recursively until the root node is reached. Therefore, the maximal number of steps is given by the number of inner nodes of the full tree. It remains to check how expensive the evaluation of $R(T)$ is and how expensive it is to find the minimal $g(t)$.

For the resubstitution estimate of $R(T)$, corresponding values for $R(T_t)$ can be calculated by passing all cases of the training set down the full tree and updating values of $R(T_t)$ for every inner node. This process has complexity $O(hN)$ as observed in the last section. The evaluation of $R(T_t)$ then takes a fixed time. Finally, finding the weakest link in each step takes at most $N_t(T)$ evaluations, so the entire generation of a pruned sequence has complexity $O(hN + |T^{term}|)$.

The final step is to find the optimal tree from the sequence. This cannot be done using the resubstitution estimate $R(T)$, because this would correspond to assuming the full tree fits the training data best.

Usually, cross-validation is used for creating a misclassification rate estimate, but that is computationally expensive. The motivation for the use of cross-validation is to compensate for small numbers of observations.

In a data-rich situation, however, it is feasible to split the dataset into training and test parts. The test set can then be used to calculate a misclassification rate estimate by passing the test data down the tree and calculating $R(T_t)$ for every inner node. Given N_{test} test observations, the process has complexity $O(hN_{test})$.

Given a reasonable height of a tree, the pruning process is directly proportional to the number cases in the dataset. The complexity is almost of the same order as the growing process except for the missing p factor. It is clear that tree models with cost-complexity pruning using test sample estimates are computationally feasible even for large datasets.

Before discussing issues encountered in practical applications of tree models for large datasets, the next section briefly considers the behaviour of statistical tests when there are many observations.

9.2.3 Statistical Tests and Large Datasets

Some pruning methods such as the Statistical Testing Procedure (STP) (Cappelli et al.; 2002) use hypothesis tests for determining "optimal" trees. The computational aspect of the application of statistical tests for pruning is dominated by the calculation of the test statistics for each node to be tested. For the most common "weakest link" approach, each iteration requires consideration of those inner nodes with terminal nodes as children. This implies that at most N_i statistics must be calculated where N_i defines the number of inner nodes. The number of cases used for the calculation is constant across one level, as for cost-complexity pruning.

In the latter case it was argued that only a constant time is needed for evaluation of an individual misclassification estimate. For a test statistic the computational complexity depends on the statistic used. The statistical testing method referred to above uses a weighted criterion of impurity. In order to satisfy independence assumptions, it must be applied to a test set, consequently the calculated criterion from the building process cannot be used. Fortunately, the statistic is additive with respect to inner nodes, so it is sufficient to calculate the partial statistic for each inner node once.

The actual testing requires the calculation of the critical value. Given the Gini Index as the impurity criterion, this is trivial as the statistics are χ^2_{C-1} distributed with C denoting the number of classes in the response variable. The handling of regression trees is slightly more complex as this requires the computation of critical values for F statistics with degrees of freedom that also depend on the number of cases in a node. In both cases the computation of the statistics is the dominating factor with $O(hN_{test})$.

Using statistical tests in conjunction with large datasets requires consideration of the computational aspect of the tests but also of the statistical aspects. Besides the usual issue of multiple testing, there is also the effect of sample size influencing the behaviour of the statistic.

The STP statistic for a classification tree T can be expressed as follows:

$$C \frac{1}{|T^{inner}|} \sum_{t \in T^{inner}} \Delta i_t \, N_t \sim \chi^2_{C-1},$$

where $|T^{inner}|$ denotes the number of inner nodes and Δi_t the impurity gain in the node t with respect to the Gini Index. To check the behaviour of the statistic with respect to the number of cases N, consider fixed class proportions in the nodes and varying numbers of cases. Because Δi depends only on the class proportions in inner nodes, it is fixed for all N the same way as C is fixed. Given trees with the same node structure and class proportions, their STP statistic will increase proportionally to N. Because the distribution is fixed and depends only on C, but not on N, increasing N will lead to the null hypothesis being rejected more often. There always exists a sufficiently large N_{p_c} such that, for any given class proportion $p_c \in (0, 1)$ of the class c, the null hypothesis will be rejected, which corresponds to retaining the split. Effectively, for large N, this means that splits containing many cases are very unlikely to be pruned, no matter how "poor" the split was in terms of impurity.

The dependency of statistical test results on dataset size has been illustrated using one practical example. This is not a special case, but rather the rule. Hypotheses tend to be rejected more often with increasing numbers of observations. Hand et al. (2000) describe several such problems encountered in practice. For large datasets, regular stopping rules based on relative impurity gain will abort the tree building earlier than test-based pruning strategies. Therefore, in most cases it is better and more efficient to use pruning methods based on misclassification or on residuals.

9.2.4 Using Trees for Large Datasets in Practice

Because tree growing and pruning procedures have linear computational costs, they are feasible for large datasets. In fact, the growing algorithm can be easily parallelized, allowing even faster computation.

One level of parallelization is possible by assigning the task of finding the best split of a variable to a single processing unit. This allows the use of p parallel units. Further parallelization is possible at node level. If the stopping rule is not met, two new nodes are created and each can be processed by an independent set of processing units. The evaluation of trees is also highly parallel because every passed case is independent of all others.

This is a nice theoretical result, but the question is how well current implementations of tree model construction cope with large datasets. Probably the most popular methods for handling trees are the rpart and tree libraries for the R/S-plus statistical environments. R version 2.2.0

has been evaluated here. As a representative of commercial packages, See5/C5.0 and its free companion C4.5 have been chosen, because they claim to be suitable for large datasets, although they are restricted to classification trees.

In order to test the performance and practical value of the programs, some real-life data were used — the US census data, extracted from the Integrated Public Use Microdata Series (IPUMS) database (Ruggles and Sobek; 1997). The complete dataset consists of approximately 25 million cases. Each case has 13 attributes of various types: continuous, ordinal, and nominal. A summary is given in Table 9.1. C4.5 and See5 expect a csv (comma separated values) ASCII data format so files of that format have been used for all tests to assure comparability.

R is a general-purpose tool, and the main memory overhead is due to data loading. Once the data are loaded in R, the tree generation itself is fast. Using the tree library, the tree construction took 9 seconds, its more flexible counterpart rpart took 47 seconds without cross-validation, both on 250,000 cases. This process consumed hardly any memory.

Both C4.5 and See5 are very fast in loading the dataset and use very little memory overhead. It is evident that these are one-shot programs specialized for the task. Nevertheless for tree generation C4.5 turned out to scale very badly for large datasets. Where C5.0 needed only 32 seconds,

Table 9.1. *Attributes of the IPUMS Census Dataset Used for the Analysis*

Attribute	Description	Type
stateicp	US State (ICPSR code)	ordinal*
ownershg	Ownership of dwelling	categorical
mortgage	Mortgage status	categorical
value	House or property value	continuous
nchildren	Number of own children in the household	ordinal
age	Age	continuous
sex	Sex	categorical
raceg	Race	categorical
marst	Marital status	categorical
educ99	Educational attainment	ordinal
empstatd	Employment status	categorical
inctot	Total personal income	continuous
migrat5g	Migration status, 5 years	categorical

*State code is treated as ordinal, because the codes are approximately geographically ordered from East to West, which makes it easier to depict effects that are on a larger geographical scale.

Table 9.2. *Performance of See5 Constructing Classification Trees of Different Sizes on the Same Dataset (the number of cases is given in thousands)*

Cases (k)	166	332	499	665	831	998	1164	1330	1496
Time (s)	31	90	95	163	217	304	432	448	484
N_t	1016	1272	1921	2121	2740	3193	3465	3865	4089

C4.5 took 429 seconds. See5 performs very well and comes close to the theoretical result of linear scalability for the construction and pruning of trees, shown in Table 9.2.

Before discussing the size of a tree, it is worth looking at datasets where many attributes and only a few observations are involved. One of the natural advantages of trees is that tree models automatically select a subset of predictors, which contributes most to the explanation of the data, because at each step a locally optimal variable is used for the split. This behaviour is especially helpful in cases where the goal is to choose best predictors from a large set, most of which contain noise. On the other hand, trees are not suitable for cases where some combination of all co-variates is likely to explain the response variable. Due to their stepwise orthogonal approach only very large trees are likely to obtain a sensible result at all, but then interpretation is difficult. Combined methods such as double-bagging (Hothorn and Lausen; 2003) can then be used.

The size of the tree varies both with the program used and the dataset size. *tree* and *rpart* use a stopping rule specifying the minimal relative improvement of a split. If the measure of impurity does not decrease by a specified percentage, the tree growth is stopped. The default setting of 1% resulted in small trees of only 4 terminal nodes, even without pruning.

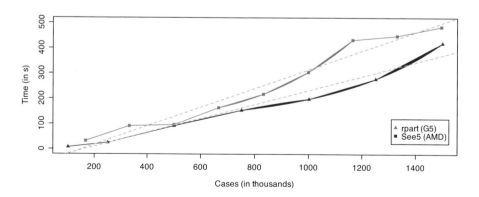

Fig. 9.1. *Scalability of* rpart *and* See5 *with increasing sample size. Different machines had to be used for licensing reasons.*

On the contrary C4.5 and See5 tend to grow very large trees. As Table 9.2 shows, the size of trees generated by See5 grows with the number of cases. However, the misclassification rate achieved was consistently around 20%. Interestingly, the misclassification rate of smaller trees was not worse, which shows that larger trees were not necessarily better.

The size of the ideal tree depends primarily on the information contained in the data and not on the dataset size. Setting a low threshold for stopping rules based on information content will produce larger trees, but in general they will not be very useful in practice unless the underlying data have the corresponding complex structure.

The most common stopping rules use an impurity decrease threshold and a node size threshold. The first rule says that the tree growth is stopped if the split does not reduce the measure of impurity by at least a specified amount. The second rule prevents splitting of nodes smaller than a given size. The importance of each rule changes with dataset size. For small datasets, the node size threshold is likely to be critical first.

This stopping rule is usually not relevant for large datasets, because the resulting nodes contain many cases. It is the impurity decrease rule that becomes critical first. The threshold must be low enough to produce prunable trees. On the other hand, too low a threshold results in late stopping, trees that are too large, and increasing computational costs.

For problems where prediction power is of primary interest and modelling data structure is less important, large trees can be used if they are carefully pruned to prevent overfitting. With large datasets there are quite enough observations for both training and test datasets.

For exploratory goals, that is discovering a structure behind the data, it is recommended to grow smaller trees, but to check their stability. If small changes in the data result in apparently randomly different trees, then the tree is not likely to be reliable. Graphical analysis of tree stability, variable importance, and other properties of a tree are discussed further in Section 9.4. Once a stable tree has been found, a graphical representation is needed to understand both the model and its relation to the data. Therefore, it is important to consider different ways of looking at the tree and to discuss questions that arise from the use of large datasets.

9.3 Visualization of Large Trees

9.3.1 Hierarchical Plots

Visualization of the hierarchical structure of a tree is the most natural way trees can be displayed. Although the idea is the same for all such plots, the placement of nodes and the representation of nodes and edges can vary. A sample plot of a tree produced from the IPUMS data is given in Figure 9.2.

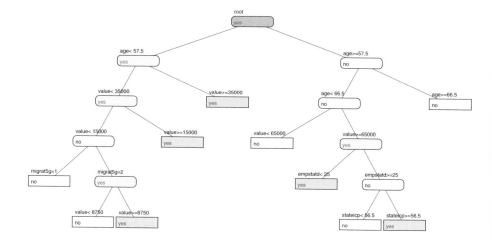

Fig. 9.2. *Classification tree based on the IPUMS data.*

The task was to predict whether the house in which the family lived was owned free and clear or whether there was a mortgage or other kind of debt on the house. The resulting tree allows some quite interesting conclusions. Younger people usually did not own the house clear unless it was of low value. Even then it was important whether they had moved into the house more recently (*migrat5g=2*), in which case they were more likely to need more money. Older people usually had no mortgage on their houses, unless the house was expensive. For expensive houses it was also important whether the owners were still employed (*empstatd<25*). Retired people were more likely to own their own house without any mortgage.

Although 400,000 cases were used for construction, the tree itself is not large. The display of the tree is no more complicated than displaying a tree based on 100 cases. Slight delays may occur if linked highlighting is used for interaction. It is necessary to identify cases in each node to determine the number of selected cases in each node. This operation is of order $O(hN)$. Selecting cases in a node is very fast, because it is possible to keep track of all case IDs present in each node.

Most other interactive operations on a tree, such as pruning, moving branches, or changing the representation of nodes, do not require access at case level and are therefore not affected by the number of cases.

There is an additional problem for large datasets that is not of computational origin but that is common for many plots. The resolution of the screen is limited and therefore causes discretization of the information presented. If each node is represented by a rectangle of width 100 pixels and the highlighting is done horizontally, then only changes of 1% or more of the displayed amount are visible. This is no issue if the data consists of around 100 cases, but for large datasets this can pose a serious problem.

This effect is amplified if absolute scaling is used, that is when the size of each node is drawn to represent the number of cases in that node. Any node of a size of 1% or less will simply disappear if displayed like this, and that is still as much as 10,000 cases for a dataset of 1 million. This problem is not specific to tree displays. Hofmann has introduced so-called "redmarking" for interactive plots in her software MANET. If a bar in a barchart or histogram would be displayed with a height of zero due to rounding, but there is at least one case to be represented, then a red line below the bar is drawn to denote the fact. Redmarking is discussed more generally in Section 4.3.7.

The sizes of nodes can be changed interactively for tree displays to counteract this effect, but nodes that are too large will overlap and overload the plot. A combination of redmarking and censored zooming can be used. Nodes larger than a certain size are displayed at a fixed size and marked as such to indicate that cropping has occurred. Redmarking is applied to very small nodes, that is they are displayed at a constant size to prevent them "disappearing". The "window" of sizes which are displayed in true proportions can be changed interactively to concentrate on nodes of interest. The idea is illustrated in Figure 9.3.

For large datasets, you can grow virtually arbitrarily large trees, if the stopping and pruning criteria are defined loosely enough. The construction of large trees is usually not desired for exploratory analyses, but it can be of interest if prediction accuracy is the main goal and overfitting does not occur. Although the interpretation of large trees is difficult, it is helpful to visualize such trees to get a global view of the tree and to check for any dependencies that may contradict known facts. Related visualization problems arise for networks with large numbers of nodes as described in Chapter 8.

The first difficulty is the appropriate placement of nodes. The nodes should not overlap, connecting lines should not cross and the order of branches should be maintained. The order of the branches is usually well defined for continuous and ordinal variables, reflecting the order of the variable or the predicted value.

One design satisfying all those requirements is the recursive bottom-up layout. The space available for the tree is divided into N_t equal parts and each terminal node is placed in the center of that space. Each parent node is placed at the center of the space used by its child nodes. This assures that no nodes overlap, because any child node can have only one parent node. The connecting lines will not cross if child nodes of the same parent are always kept together. The placement in the height-axis can be arbitrary as long as child nodes are always plotted below their parent. The possibilities include equidistant partitioning of the height space or

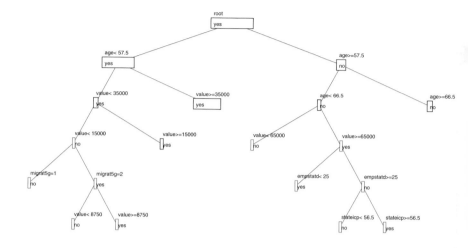

original node size

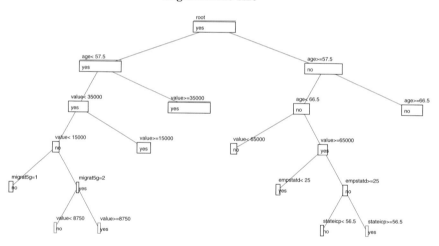

4x node size zoom with censoring

Fig. 9.3. *Censored node size zoom illustrated on a classification tree with node sizes proportional to the numbers of cases. Nodes represented by a red rectangle are in fact smaller than the displayed size; nodes with a red bar atop are in fact larger.*

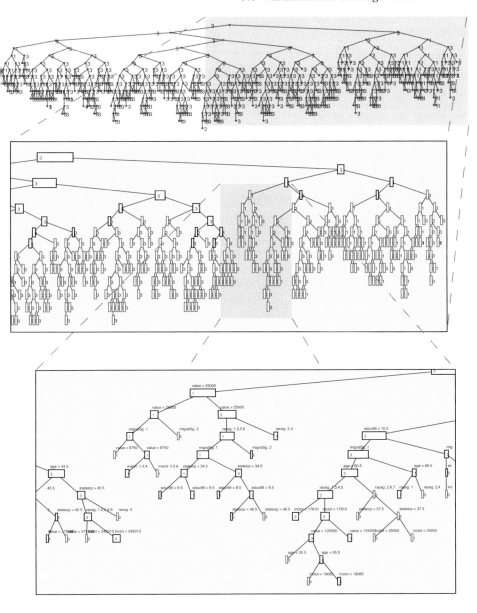

Fig. 9.4. *Logical zoom for large trees, illustrated on an unpruned tree based on the IPUMS data.*

using $h\sqrt{\frac{n_t}{N}}$, where n_t denotes the number of cases in node t. This layout is used by KLIMT for its flexibility and is suitable for small and medium sized trees.

Many different layouts exist in order to reduce the space used by the graph. It is partly a matter of taste, because most layout methods try to include an aesthetic optimization, but experience suggests that reducing the space used makes the plot harder to interpret, because terminal nodes are no longer evident at a glance.

Consider the size of a tree that can be displayed on a screen in its entirety. Given a node size of 100×40 pixels (including the surrounding free space), a working window of 800×600 allows a maximum of 8 visible nodes at one level for vertical orientation and 12 for horizontal orientation. This is clearly inadequate for large trees.

One solution is the use of interactive pan and zoom. To get an overview, the entire display can be zoomed out to present a structural overview. Local analysis can be done by zooming into the area of interest and exploring in the local neighborhood. Logical zooming (see also Section 4.4.3) can be used to prevent overloading as illustrated in Figure 9.4. At fixed zoom threshold levels, the display changes from more detailed to less detailed and vice versa, similar to the way many electronic maps are implemented nowadays.

The use of multiple views of the same tree simplifies navigation and improves interpretability. One smaller overview window with large zoom out gives a bird's eye view and can be used to capture the overall structure while another bigger working window is used for further zooming in.

Alternative techniques proposed for large hierarchies are the spherical projection hyperbolic browser (Lamping and Rao; 1995) and fisheye views (Furnas; 1986). In all cases, the original planar layout is projected in a non-linear way. The idea is to emphasize an area of interest by using larger distances between nodes in the projection center than for nodes further away. The main problem of all distorting projections is navigation in the space. Because of the distortion, structures that were identified in one view can seemingly fall apart or be hard to locate again when the projection changes. There are several layout libraries available for this task, which are used in computer science, but such techniques have not been used for statistical trees yet.

9.3.2 Sectioned Scatterplots

Statistical trees can be seen in more ways than just as a hierarchical structure. In their role as a predictor, they partition the space of covariates, usually orthogonally to the axes. For each partition, a prediction model is used. In the case of CART this is a constant model. A natural way to explore the impact of continuous and ordinal covariates on the tree is

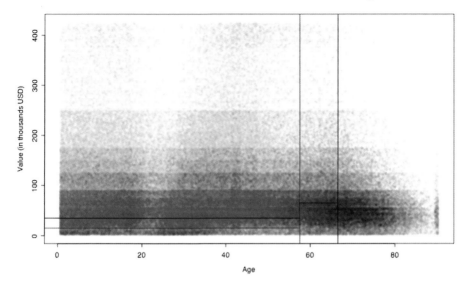

Fig. 9.5. *Sectioned scatterplot of property value vs. age. Black lines in the plot represent splits. Red points represent houses with some kind of mortgage.*

to draw sectioned scatterplots. Each scatterplot is a two-dimensional projection of the covariate space and therefore it can be enhanced by adding sectioning lines orthogonal to the axes and corresponding to the splits.

Figure 9.5 shows a sectioned scatterplot. It is a textured scatterplot[4] that displays the value of the property and the age of the family member for more than 600,000 cases. α-blending is used to prevent overplotting and provide a rough density estimator. It is evident how the model attempts to separate people with low-value houses, which are unlikely to be mortgaged, and older poeple that are likely to have paid their mortgage in full already.

The primary drawback of sectioned scatterplots is that only two covariates can be shown at once and so splits on other variables are not visible. General solutions for this problem offer matrices of sectioned scatterplots or interactive methods, where only a subset of the total population is displayed. Especially in the interactive context, sectioned scatterplots provide a nice way of analysing trees, because they visualize the neighborhood of a split, allowing a visual assessment of the quality and variability of each split. Most interactive tools available for scatterplots, such as pan, zoom, brushing, or queries, apply to sectioned scatterplots as well.

Are sectioned scatterplots suitable for large datasets? Trees tend to use only a limited set of covariates even though the dataset may consist

---

[4] A textured scatterplot is the generalization of a textured dotplot (basically a more sophisticated form of jittering) as described in Tukey and Tukey (1990).

of many attributes, as discussed in Section 9.2.4. Such a limited dataset may not be large, but it usually contains more than two covariates.

One complementary method to obtain more informative sectioned scatterplots is to use a technique similar to node size censoring described in the previous section. The idea is to plot a sectioned scatterplot for a subtree instead of for the entire tree and to stop after a certain depth of the tree. The resulting plot concentrates on a local neighborhood of the subtree relative to the hierarchical structure.

Selecting a subtree corresponds to zooming into the original plot, if the tree contains only two covariates. In other cases, selecting a subtree also leads to a change of projection, because other covariates will be used. This process can be repeated recursively to allow a complete walk through the tree, though the changes of projection must be treated with care. Guidance can be provided by a hierarchical plot displayed in parallel.

The use of large datasets leads also to problems inherent in scatterplots in general, as discussed in Section 3.4.2. A cloud consisting of many observations results in a continuous monochromatic area. Any lines shown in such areas delineating splits are uninformative, because the presence of many points hides any structure that the tree tries to depict. Many different techniques have been used to solve this problem, because it is not unique to sectioned scatterplots. In most cases, some kind of density estimate is used.

Most density estimation methods are suitable for sectioned scatterplots, but there are some points to consider. Due to the orthogonal nature of splits, methods using different geometry such as hexagonal binning cannot be recommended. Rectangular binning gives good results. It is possible to modify the range and binwidth in order to match the binning to the split under inspection. A slight variation of the bin range can reveal extra information about the splits, such as whether they are sharp or not.

Interactive case selection can be a problem if density estimates are used. If only a few cases are selected, they can be directly plotted on top of the scatterplot, but with large datasets this is rarely the case. For large selections, it is necessary to combine density estimates for the data and for the selected part, possibly using α-blending.

In many cases, continuous variables are only available in discretised forms. A textured version of the scatterplot in Figure 9.5 has been used, because the underlying data consist of relatively few discrete values. Age is given in years and property values in thousands. When the levels for both variables are aggregated, it is possible to use fluctuation diagrams as for categorical data, but with levels ordered according to their value. It is possible to draw sections in fluctuation diagrams in the same way as they are plotted in scatterplots. An example of a sectioned fluctuation diagram is given in Figure 9.6.

Fluctuation diagrams provide a density estimate and also allow the use of linked highlighting. In Figure 9.6, all houses without mortgages

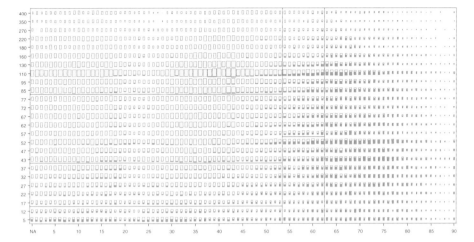

Fig. 9.6. *Sectioned fluctuation diagram of two discretized continuous variables* age *and* value *with highlighted cases where the family owns the house free and clear.*

have been highlighted in red. The plot shows both the approximation of the bivariate distribution of all cases by the size of each rectangle, and the conditional distribution of the selected cases. It reflects the structure as described by the tree: older people and families owning cheap houses tend to be clear of mortgage.

Interesting information about the dataset can be derived from the fact that even children are registered as owning houses. All members of the family are recorded in the database with the attributes of the entire family. This explains the high-density area among children up to 18 years of age and the gap afterwards. Young people older than 18 years of age usually leave the family and rent, so they do not appear in the dataset until they buy their own house.

Fluctuation diagrams are area-based plots, unlike scatterplots. This makes them practical for dealing with large datasets. The sectioned scatterplot in Figure 9.5 takes minutes to draw, whereas the fluctuation diagram in Figure 9.6 is displayed within a few milliseconds. There are more area-based plots that can be used for the representation of trees. The advantages of area-based plots for large data are discussed in Section 3.3.

9.3.3 Recursive Plots

A common way of handling large datasets is to use area-based plots instead of case-based plots. It is possible to partition a fixed rectangular area in the same way the tree partitions the sample data. This idea is utilized by treemaps. A treemap is constructed by partitioning the available plot space alternately horizontally and vertically according to the proportions of cases belonging to each branch. An additional enhancement in the form of depth-shading has been included. Each partition is shaded according to the depth of the associated leaf. Leaves that are the result of more splits are deeper down and therefore darker. This helps to display the structure of the tree. A shaded treemap is shown in Figure 9.7.

Treemaps are independent of the number of covariates used in the splits and they also do not depend on the number of observations, so they are suitable for large datasets.

The number of cells corresponds to the number of terminal nodes, so the behaviour of treemaps is not determined by the number of cases, but by the size of the tree. As with the visualization of hierarchical structures, the problem of treemaps for large trees has arisen in computer science as well (Turo and Johnson; 1992).

Many problems result from the fact that discretization is necessary in order to construct a picture on a pixel-based medium, such as a computer screen or printer. Due to screen resolution constraints, nodes that are too small may not be drawn if their correct width or height is less than one pixel. This problem is usually addressed by specifying a minimal width and height of one pixel and using a different colour to denote non-proportional representation. This approach leads to another problem, if many small nodes are present. Assume a tree is split in 11 branches where one branch includes 50% of all cases and the remaining have 5% each. Given 10 pixels for the entire branch, the ideal distribution would be to use 5 pixels for the first branch and 0.5 each for the remaining ten. Using a single pixel for each of the 10% branches ensures that there is no space at all for the first and largest branch. Therefore for every treemap

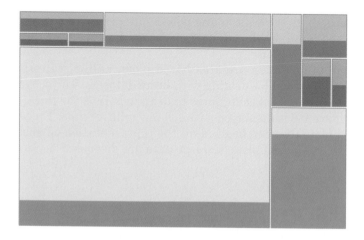

Fig. 9.7. *Depth-shaded treemap. Selected in green are cases with a mortgage.*

there is a minimal size that is needed for the display. If the screen size is smaller than this, zoom and pan must be used for navigation and logical zooming to provide an overview-treemap.

Another issue arises from the fact that each partition consists of a border and an inner area. A partition of size 2 pixels or less cannot have both, because each border needs one pixel and yet another pixel is needed to represent the inner area. In a truly proportional setting, a border belongs to at least one partition. In practice, there are two ways of plotting a recursive area-based plot. The first method is to reduce the number of available pixels by the number of fixed pixels, such as borders, and to partition the remaining space according to the relative size of each partition. The second method is to take the total available area, determine the exact position of each partition and draw a line at the rounded exact position. The latter method results in a proportionately correct display, but can lead to partitions disappearing. The former method can produce the overfilling effect described in the previous paragraph.

Both effects can be circumvented by using zoom and pan. Labeling of treemaps is a problem due to space constraints, so interactive queries offer the most appealing solution.

Treemaps are not negatively affected by the number of cases since they are area-based plots. Problems arise if the size of terminal nodes varies very much or if too many small nodes are present. These problems are inherent in all area-based plots and can be solved by using interactive navigation as discussed in Section 4.4.3. Spineplots of leaves (e.g., Figure 9.8) are constructed in the same way as treemaps, except that the splitting direction remains fixed instead of alternating. Treemaps and spineplots share the same properties with respect to large datasets. Both plots are suitable, given the support of interactive techniques.

Fig. 9.8. *Depth-shaded spineplot of leaves with cases with a mortgage selected. The order of bars corresponds to the order of terminal nodes in Figure 9.2 from left to right.*

9.4 Forests for Large Datasets

The idea of classification and regression trees has been around for some time. Trees have proved to be powerful for non-linear problems and for cases where weak or no assumptions about distributions of the covariates can be made. These properties meant that trees were used for more than just exploratory purposes, where their good interpretability was the main advantage. This has led to research on improvement of the prediction accuracy of tree models.

The aim is to produce better classifiers by combining predictions from several trees. Many methods have been suggested including bagging (Breiman; 1996), boosting (Schapire; 1999), and random forests (Breiman; 1999). Unfortunately, forests are often impenetrable, because the nice interpretability of a single tree cannot be extended to an entire forest. There is indeed information hidden in the forests and graphical methods for extracting it have been proposed, especially for forests generated from large datasets. Apart from information about prediction accuracy, there is information on the stability of trees (how different the trees in the forest are), and there is information on the covariates used. Given a large set of covariates, trees tend to use only a few for splitting. Graphics can help to show the contribution of each predictor variable to the models.

In the following, the analysis is restricted to classification trees, but the results and methods can be extended easily to regression trees. With large datasets, it is possible to construct many trees very easily by taking random samples from the dataset. The samples can be as large as 90% of the training data. Small dataset problems, which need methods like bootstrapping, are not relevant for large datasets.

R was used for the analysis and the sample size was limited to 250,000 cases. For each tree, 500,000 cases were sampled and split into two equal sets, one for training, and one for testing. In all, 200 trees were fit to the training data. The generation of a single tree took approximately 20 seconds so the construction of forests even for large datasets is just a matter of minutes.

In order to analyse the trees as a group, "forest-data" consisting of information and statistics for each inner node in every tree were calculated. The information included the name of the tree, the node ID,[5] the variable used for the split in the node, the number of cases of the test set falling into that node, and the deviance gain of the node based on the test set. Interactive visualization tools can be used to evaluate these data.

---

[5] In a full binary tree, the number of nodes in each level is fixed, therefore it is possible to give each node a unique ID, usually starting from the root, then left to right at each level. Every binary tree can be seen as a pruned full binary tree. Therefore, it is possible to define unique node IDs for every binary tree. R produces binary trees.

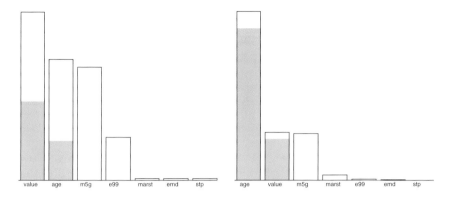

Fig. 9.9. *Use of variables. Barchart of the frequency of variables (left). Weighted barchart of the variables used with deviance gain as weight (right). Highlighting corresponds to the first two splits in a tree.*

First, take a closer look at the variables used. The left plot in Figure 9.9 shows a barchart of how often variables were used. The height of each bar corresponds to the number of nodes that contained a rule featuring the corresponding variable. The highlighted part corresponds to the first two splits of a tree. It is evident from the plot that value was the most frequently used variable, followed by age, migrat5g, and educ99.

This plot does not necessarily show the importance of each variable. If a variable is used quite often for local partitioning of a few cases in a late split, then its frequency will be high, but not its importance. A weighted barchart with deviance gain as weight, as shown in the right plot of Figure 9.9, displays the relative importance of variables better. From that plot, it can be concluded that age explains more information contained in the data than value, although it is not used as frequently. Migration appears to be quite important, although it is used more in lower splits as the highlighting shows.

Both plots display aggregated information obtained from all trees. Another interesting question is how different the generated trees are. In order to answer that question all splits should be visualized. One way to do so is to use a "Tetris plot" as shown in Figure 9.10. Each coloured box corresponds to one inner node, and each splitting variable has a specific colour. The plot reveals that the trees are very similar for the first three levels. Even further down the tree, their differences are rather in the order of the variables used, not in their type.

The usage of variables in each tree can be checked using a fluctuation diagram of trees and variables weighted by deviance gain (Figure 9.11). The plot confirms that the trees are very similar. This plot would show differences in the behaviour of variables across trees if there were any.

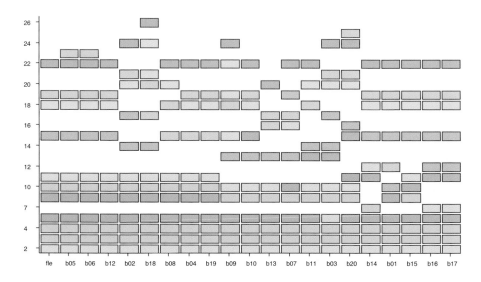

Fig. 9.10. *"Tetris" plot. Each rectangle corresponds to a node. Trees are plotted on the X axis, node IDs on the Y axis. Each colour represents one variable.*

If the structure of the trees appears similar, then it is likely that the performance of the trees on training data will also be similar. This can be checked by looking at the misclassified cases. 250,000 test cases that were not present in any training set have been passed down all generated trees. Two plots in Figure 9.12 summarize the results and confirm the supposition.

The left plot shows how often each case was misclassified. The left-most bar corresponds to zero, hence it consists of cases that were always correctly classified. The rightmost bar consists of cases that were always misclassified. The sharp separation is very unusual for a forest of trees for small data. Usually there are many cases misclassified a couple of times, and cases that are always misclassified are rare. The plots show that an improvement of the model by aggregation cannot be achieved.

Cases having a mortgage are highlighted in the plot. By looking at the highlighting in the rightmost bar, it is obvious that the models tend to misclassify cases having no mortgage far more often. This can be verified easily by plotting bars for misclassified cases in each tree as in the right plot of Figure 9.12. The total misclassification rate of the trees varies only slightly, the misclassification rate of the individual classes shows no big variation either.

Another 200 trees were grown, but they only confirmed that the trees based on this large dataset are very stable, and their accuracy cannot be enhanced by aggregation methods. All the proposed plots scale well

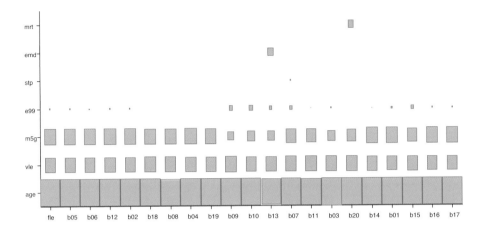

Fig. 9.11. *Weighted fluctuation diagram of trees (x-axis) and variables (y-axis) with deviance gain as weight.*

for large datasets and all but three (the "Tetris" plot, the tree-variable fluctuation diagram, and the misclassification plot) are not dependent on the number of trees. For these three plots, 21 trees have been displayed for illustration. The plots are useful for up to about 200 trees. None of the large datasets analysed so far have produced sufficiently different trees to warrant growing more. On the contrary, the datasets that benefited most from large forests were only of several hundred cases.

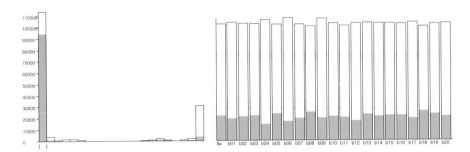

Fig. 9.12. *Histogram of the number of models misclassifying a case (left). The misclassification plot (right) consists of bars representing misclassified cases for each tree. The cases having a mortgage are highlighted (green) in both plots.*

9.5 Summary

Trees are very useful for both exploratory and predictive tasks. In theory, tree models can be applied to large datasets as well as to small. In practice, the freely available software is not yet suitable for generating trees for very large datasets, but some specialized commercial packages prove that it is possible. Modification of the existing code-base to support large datasets and incorporating parallel processing should not be too demanding, but it still needs to be done. Effective tree visualization is dependent on having efficient and fast model-fitting algorithms.

Some implementation techniques, aimed especially at faster construction of trees, have been considered by researchers in Data Mining as discussed by Gehrke et al. (2000).

Visualization of large hierarchical structures has been a research topic in computer science for quite a while. Unfortunately, little has been done so far for statistical trees. Large datasets do not necessarily lead to large trees, but several issues arise when combining large datasets and trees of unlimited size. Interactive methods can be used to help to solve some of the problems.

A statistical tree is not just a hierarchical structure, but can be also viewed from other aspects. This chapter has discussed the behaviour and scalability of several alternative visualization methods such as sectioned scatterplots, fluctuation diagrams, treemaps and spineplots of leaves for large datasets.

Large forests of trees can be grown to assess the stability of individual trees, to estimate the importance of individual variables, and to improve the prediction power of the model. Several visualization tools have been described that help to understand the behaviour of the trees and to get more information from datasets.

10

Transactions

Bárbara González-Arévalo, Félix Hernández-Campos, Steve Marron, and Cheolwoo Park

'Cause there's a million ways to go, you know that there are...

Cat Stevens, *If you want to sing out, sing out*

10.1 Introduction and Background

The area of internet traffic measurement and modelling has a pressing need for novel and creative visualization ideas. The issues and the data are both complex, yet few researchers in the area (with some notable exceptions) are aware of the power of visualization for addressing the problems and understanding complicated behaviour.

The internet shares some similarities with the telephone network. Both are gigantic, worldwide networks for the transmission of information. Both share the notion of "connection", generally between two points. For this reason, the first models for internet traffic were based on standard queueing theory, with assumption of Poisson arrival of connections, and of exponentially distributed times of connection duration.

A large body of exciting work during the 1990s revealed that these assumptions were grossly inadequate, and far different models were usually much more appropriate. In particular, duration distributions exhibit heavy tails (caused by both far shorter and also far longer connections than are typically found in telephone traffic), and time series of aggregated traffic exhibit bursty behaviour and long range dependence. An elegant mathematical theory demonstrating how heavy tail durations can lead to long range dependence was developed by Mandelbrot (1969), Cox (1984), Taqqu and Levy (1986), Leland et al. (1994), Crovella and Bestavros (1996), Heath et al. (1998), and Resnick and Samorodnitsky (1999).

This theory is deep and compelling and gives good descriptions of observed behaviour in a wide range of circumstances. However, there has been recent controversy at several points.

One controversial point has been the issue of the heaviness of tails. Downey (2000) suggests that the Log Normal (not heavy tailed in the classical sense) can fit duration distributions as well as classical heavy tail distributions and gives some interesting physical motivation for this distribution as well. However, by developing the nice idea of tail fragility, Gong et al. (2001) showed that both types of distribution can give apparently reasonable fits. This general direction was further developed by Hernández-Campos et al. (2002) using much larger data sets (in the millions), together with a novel visualization for understanding the level of sample variation. This latter work showed that a mixture of three Double Pareto Log Normal distributions (Reed and Jorgensen; 2004) gave an excellent fit and was also physically interpretable. These results motivated the development of the concept of variable heavy tails.

Another point of controversy has been the issue of long range dependence. This is currently widely accepted (and intuitively sensible), but some interesting questions have been raised (using some novel visualization ideas) by Cao, Cleveland, Lin and Sun (2001, 2002a,b,c).[1] The key idea is that aggregated traffic, of the type typically found at major internet nodes, tends to wash out long range dependence. The idea is theoretically justified by appealing to limit theorems for aggregated point processes, and supported by recent measurement studies that examined high-capacity backbone links, see Zhang et al. (2003) and Karagiannis et al. (2004). An example, where both types of behaviour were observed, depending on scale, was studied using some different visualizations, by Hannig et al. (2001). An interesting issue to follow in the future will be the state of this balance between long range dependence caused by relatively few but extremely large transmissions, and a more Poisson type probability structure caused by aggregation. Cao, Cleveland, Lin and Sun (2001, 2002a,b,c) predict ultimate Poisson type structure, for the good reason that internet traffic is continually increasing. However, this is based on an assumption that the distribution of sizes of transmissions will stay fixed, which seems questionable.

Downey (2001) questioned long range dependence from a different viewpoint, by showing that duration distributions may not be very consistent with the definition of heavy tailed, in the classical asymptotic sense. This was the first observation of variable heavy tails, as defined in Hernández-Campos et al. (2002). That paper goes on to develop an asymptotic theory that parallels the classical theory. In particular it is seen that long range dependence still follows from the far broader (and more realistic in terms of the nature of the data) concept of variable heavy tails.

---

[1] For a collection of papers on internet traffic modelling, see http://cm. bell-labs.com/cm/ms/departments/sia/wsc/webpapers.html.

This paper points out some challenging visualization problems and offers approaches to their solution.

The first main problem is related to the *Mice and Elephants* graphic, developed in Marron et al. (2002), discussed in Section 10.2. The problem is how to choose a representative sample, and it is seen that the usual device of random sampling is clearly inappropriate. In Section 10.3, two different biased sampling approaches to this problem are proposed. A quite different approach to visualizing this large and complex population, based on studying quantile windows, is shown to reveal other interesting structure in the data in Section 10.4. This work on flow sampling for visualization complements existing ideas developed in the context of sampled traffic monitoring. Recent work in this area, Duffield and Lund (2003) and Hohn and Veitch (2003), has shown that flow sampling provides a more accurate view of internet traffic than packet sampling. In particular, Hohn and Veitch have shown that flow sampling preserves long range dependence.

The second main problem is motivated by an apparent "commonality of flow rates", discussed in Section 10.5. A large scatterplot seems to reveal interesting visual structure that makes physical sense. The question is how to best understand the driving phenomena.

10.2 Mice and Elephant Plots and Random Sampling

The Mice and Elephants plot is a visualization that illustrates the fundamental theory discussed in Section 10.1. In particular, it shows how a heavy tailed distribution can lead to long range dependence. An example is shown in Figure 10.1. The key idea is that internet flows, i.e., the set of packets that make up a single connection, are represented by line segments. The left (right) ends of the line segments show the times of the first (last) packets in each flow. Thus, each line segment represents the overall time of activity of that flow. For good visual separation of the line segments, the data have been jittered by adding a random height on the vertical axis. This is in contrast to the common packet arrival time vs. address plots commonly used for traffic anomaly detection, as in Cho (2001).

The data here were HTTP (Web browsing) response times. They were collected during a four hour period, 8:00 AM to 12:00 noon, on a Sunday morning in April 2001 at the University of North Carolina (UNC). This time period was chosen to represent a light traffic time. For a parallel analysis of a heavy traffic time, see Marron et al. (2002). More detailed graphics for both analyses are available at the web address: http://www.cs.unc.edu/Research/dirt/proj/marron/MiceElephants/. For more details on the data collection and processing methods, see Smith et al. (2001).

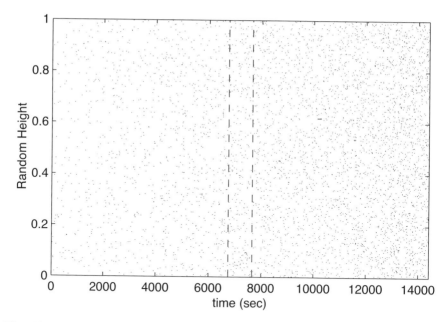

Fig. 10.1. *Mice and Elephants plot for a full four-hour time block. Vertical bars indicate the 15-minute time window shown in Figure 10.2. This suggests that all flows are "mice".*

The total number of flows for the time period in Figure 10.1 was 1,070,545. Massive overplotting resulted from an attempt to plot all of them. A simple and natural approach to the overplotting problem is to plot only a random subsample. This was done for a subsample size of 5,000 (chosen for good visual effect) in Figure 10.1, and in the other figures in this section.

Figure 10.1 shows steadily increasing traffic, which is the expected behaviour on Sunday mornings (perhaps the times at which students begin web browsing is driven by a wide range of adventures experienced on the previous night!). It also suggests that there are no long flows, with the longest visible flow being less than 5 minutes. This is a serious misimpression, which completely obscures the most important property of the traffic.

This point becomes clear from a similar graphic, zoomed into the region between the vertical bars in Figure 10.1, which represents the central 15 minutes (1/16th of the total time). Figure 10.2 shows this zoomed mice and elephants plot. There were 59,113 (not far from 1,070,545 / 16) flows that intersected this time range. Plotting all would again result in severe overplotting, so only a random sample of 5,000 is plotted.

The visual impression of Figure 10.2 is far different from that of Figure 10.1. In particular, there are a number of flows that cross the full

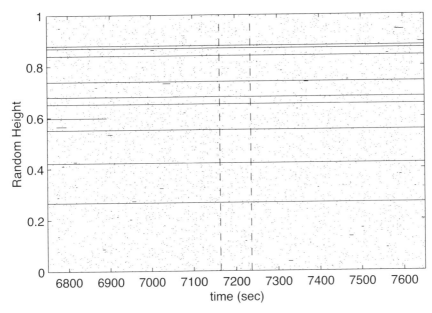

Fig. 10.2. *Mice and Elephants plot for a 15-minute time block. Vertical bars indicate the time span containing the 5,000 flows, shown in Figure 10.3. This shows both "mice" and "elephants".*

15-minute interval, which seems quite contrary to the visual impression of Figure 10.1 — where all flows appeared were much less than 5 minutes in duration. This misimpression is caused by a combination of the heavy tailed duration distribution and the random sampling process. Because of the heavy tails, there are only a very few flows that are very long. These have only a small chance of appearing in the randomly selected sample. For example, the chance that any of the biggest 40 flows have a chance of appearing is only about $(40 \cdot 5,000/1,070,545) \approx 0.19$. The number 40 is relevant, because 38 flows extend the full length of the central one hour time interval. This small probability of inclusion explains why none of these very long flows appears in Figure 10.1.

It is interesting to zoom in once again. Figure 10.3 shows the results of repeating the visualization for the region between the vertical bars in Figure 10.2. Those bars do not show 1/16th of the region in Figure 10.2, because that contains less than 5,000 flows (which would give an inconsistent visual representation). Instead, the bars are chosen so that exactly 5,000 flows intersect the time interval (which is again centered in the range of the data), which is about 1.3 minutes long.

Figure 10.3 shows a rather large number of long flows, and, because there is no sampling, is representative of behaviour at a given time. However, this view is also biased because in some sense it shows too high a

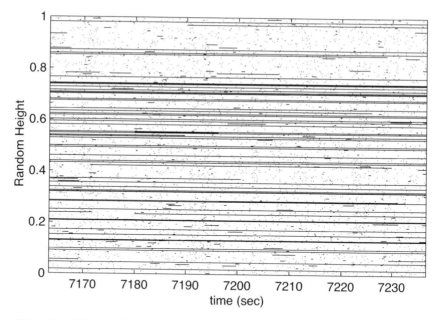

Fig. 10.3. *Mice and Elephants plot for time window containing 5,000 flows.*

proportion of long flows. The reason is a length biasing type of effect: long flows have a much greater chance of appearing in any such small interval, yet, as noted above they are a very small fraction of the population.

The clear conclusion from Figures 10.1, 10.2, and 10.3 stands in stark contrast to one of the most time-honored principles of statistics (and a commonly used tool in visualization): simple random sampling of these data does not give a representative sample. Again, this problem is caused by the heavy tails of the duration distribution of internet connections, and there is a general principle at work: simple random sampling will never give a representative sample in heavy tail situations.

The first open problem proposed in this paper is to find an improved version of a representative sample. A sensible first step may be to decide what that means. Is there a reasonable mathematical definition that makes sense for heavy tailed distributions? Can classical length biased sampling ideas perhaps be useful? Two biased sampling approaches to these problems are given in Section 10.3. The first approach essentially considers a wide range of zooming window views of the data, of which Figures 10.1, 10.2, and 10.3 are examples, and uses probabilities of appearance in the window to assign sampling weights, as discussed in Section 10.3.1. The second approach assigns weights using the Box–Cox transformation of the data, with details appearing in Section 10.3.2.

Figures 10.1, 10.2, and 10.3 show that the name *Mice and Elephants* is sensible for this graphic. It has become commonplace terminology in the internet research community for this phenomenon of a very few, very large flows. The concept is fundamental to the ideas outlined in Section 10.1. It is a clear consequence of the heavy tail duration distributions. It also makes the long range dependence in the aggregated time series visually clear. In particular, time series of binned traffic measures (such as packet counts) are essentially vertical sums of the line segments in the mice and elephants plots. The theory described above about heavy tails implying long range dependence is visually clear, given that the very long elephant flows clearly persist over long time ranges. It is not surprising that the persistence of the elephants results in the often observed bursty behaviour of internet traffic.

Another interesting open problem is to use this visualization to motivate new quantitative measures for understanding the nature of this type of data. The standard notions of heavy tails for the duration distributions, and of time series dependence, are not designed for describing the full structure of these data. Instead they are just tools adapted from other areas, which perhaps result in a somewhat clumsy statistical analysis. Can the quantitative analysis be sharpened by quantifying other aspects of the full plot?

Mice and elephants visualizations also demonstrate clearly that standard queueing theory models, with exponential duration distributions, are grossly inappropriate. This is seen in Figure 10.4, which is a duplicate of Figure 10.3, but for simulated data with exponential distributions. To keep the comparison as fair as possible, the real data time range, sample sizes and even start times are used. Only the duration of each flow (the length of the line segment) is simulated. The exponential parameter was chosen to give a population mean that was the same as the sample mean for the real data.

Figure 10.4 shows a completely different type of distribution of flow lengths from the real data shown in Figure 10.3. In particular, there are no flows that are nearly long enough to cover the whole interval, the number of very short flows is far fewer, and there are many more "medium size" flows. This is a consequence of the "light tail" property of the exponential distribution. Once the mean is specified, there are constraints on the frequencies of very large and very small observations. These constraints make the exponential distribution a poor approximation to the type of behaviour seen in Figures 10.1, 10.2, and 10.3. These mice and elephants plots support the conclusion from Figure 10.1 that classical queueing models are inappropriate for internet traffic. In addition, the mice and elephants plot in Figure 10.4 seems consistent with the idea that, when this traffic is vertically aggregated, the resulting time series exhibit only classical short range dependence.

The problem proposed above of how to subsample for effective mice and elephants visualization should not be regarded as "one off". The internet is constantly changing in many ways, and this could become a stan-

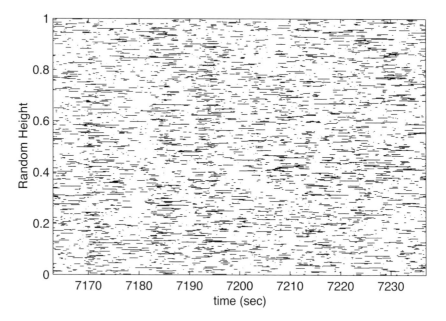

Fig. 10.4. *Mice and Elephants plot for simulated exponentials, in setting of Figure 10.3.*

dard tool for monitoring change. For example, monitoring could help to resolve the controversy as to whether large-scale aggregation will eventually swamp out long range dependence effects or whether the latter will continue with the continuing growth of elephants in frequency and size. An effective solution might also extend well beyond internet traffic, and might provide the beginnings of a new theory of sampling in heavy tail contexts.

10.3 Biased Sampling

In this section, two biased sampling methods are proposed, which generate samples that give a better visual impression of the nature of this complex population. The first is based on *windowed biased sampling*. The second takes a quite different *Box–Cox biased sampling* approach. Both approaches have connections to classical length biased ideas but are adapted to the present special setting.

Both approaches provide a visualization over the full time interval, as shown in Figure 10.1, involving a non-uniform random sample of the full set of flows. But each involves construction of a weighting scheme, that gives the elephants an appropriate probability of appearing.

In this section, only responses with nonzero duration are considered. The number of these responses is $N = 382127$.

10.3.1 Windowed Biased Sampling

Probability weights are chosen to mimic the chance of appearing in a window that is the right size among a succession of zooming windows. Figures 10.1, 10.2, and 10.3 are members of this sequence of zooming windows.

Figure 10.5 shows the result of the biased sampling method of this section. It shows nearly 5,000 (actually 4834) HTTP responses, which are drawn in a way that gives a much clearer impression of the Mice and Elephants nature of the dataset.

Notation for making this precise, which allows straightforward generalization to other contexts, is now introduced. Let L denote the total time period (4 hours for these data, as seen in Figure 10.1), and let x and d be the starting time and duration of each HTTP response, i.e., x is the left endpoint and $x + d$ the right endpoint of each line segment. Let N

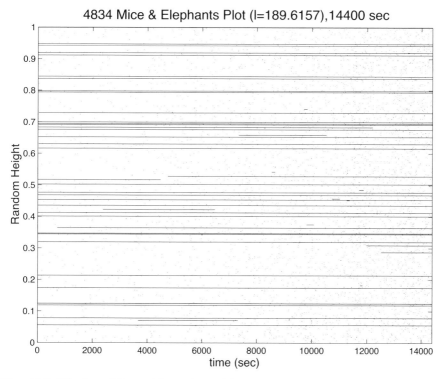

Fig. 10.5. *Windowed Biased Sampling of Sunday morning HTTP Response data. Shows both, mice and elephants.*

denote the total number of HTTP responses. Then the density of the data is $\delta = \frac{N}{L}$. Figure 10.1 shows that the density of the data is clearly not constant over time (there is a definite increasing trend), so δ should really be viewed as the average density.

Let n denote the number of responses desired to appear in the final full time scale visualization (taken to be 5,000 in Figures 10.1, 10.2, and 10.3). Let l be a candidate subwindow size (i.e., l indexes a zooming family of windows). An interesting view over a wide range of values of l is available in the Windowed Biased Sampling section of the webpage González-Arévalo et al. (2004).

An appropriate choice of l will give a good subwindow approximation of the overall average data density δ. Thus, it makes sense to use $l = \frac{n}{\delta}$.

The HTTP responses that appear in the final full time span visualization — shown in Figure 10.5 — are chosen with a weighted random sampling scheme, where the weights are chosen to reflect the probability that the given flow appears in a window of length l.

To calculate these weights, consider a point U chosen at random from the interval $[0, L - l]$. Then all the responses in the interval $[U, U + l]$ could be taken as the sample. In this case, the sample size will be close to n. But this sample has the drawback that it does not cover the full time range $[0, L]$. To cover the full range, a weighted random sampling scheme has been devised as follows.

Because some responses begin before time 0, or finish after time L, weights are recommended that depend on whether or not the response was fully captured, meaning that all data packets were transfered, within the interval $[0, L]$. HTTP responses that had some of their data transferred either before the starting time or after the ending time, are called censored.

First weights are assigned to the fully captured responses. The response is modelled as having a random start time X, assumed to be uniformly distributed on $[0, L - d]$. Let $p(d, l)$ be the probability that an observation, with duration d, intersects the random window $[U, U + l]$. Let $g(x, d, l)$ be the conditional probability that an observation is selected given that its starting time is x, its duration is d, for the window width l. Let $f(x) = \frac{1}{L-d}$ be the probability density function of X. Then

$$g(x, d, l) = Pr(U \in [x - l, x + d]) \qquad (10.1)$$
$$= \frac{\min(x + d, L - l) - \max(x - l, 0)}{L - l}.$$

Averaging this over x values, i.e. taking expectation over X, gives an approximate probability that a randomly chosen response intersects the window $[U, U + l]$,

$$p(d,l) = Eg(X,d,l) = \int_0^{L-d} g(x,d,l)f(x)dx$$

$$= \max\left\{1 - \left(1 - \frac{l}{L-d}\right)\left(1 - \frac{d}{L-l}\right), 0\right\}, \qquad (10.2)$$

which is used as the weight for random selection of responses.

Second weights are assigned to the censored responses. A similar calculation gives

$$p(d,l) = \min\left\{\frac{d}{L-l}, 1\right\}. \qquad (10.3)$$

Note that any flow with duration at least $L-l$ will have probability one of being in the sample. Thus, all 38 HTTP responses that cover essentially the full 4-hour time interval appear in Figure 10.5.

So, for fixed l and L each observation can be chosen with probability $p(d,l)$ given in (10.2) and (10.3). This will give a sample that spans the whole time interval, but that gives a higher probability to the elephants. A possible drawback of this method is that the size of the sample is random, and depends on l, although it will be close to $n = l\delta$.

When a sample of exactly size n is required, the following can be done:

1. Start with a larger window, with width $l = \frac{n}{0.8 \times \delta}$.
2. Choose each observation with probability $p(d,l)$.
3. From this larger sample, choose a random sub-sample of size n.

Note that the number 0.8 is arbitrary, and intended to guarantee that there are at least n points in the sample. However, care should be taken, in the choice of this scale factor, to ensure that the original sampling scheme produces a sample not much larger than the desired n. Otherwise, the subsequent random sampling will tend to reproduce the effect that the procedure is trying to avoid: random sampling downweights the elephants.

10.3.2 Box–Cox Biased Sampling

As discussed in Section 10.2, random sampling does not provide a useful visualization of durations of HTTP responses because of the heavy-tail of the duration distribution. Here, a biased sampling approach based upon transformations is considered. In this section, the censored observations are handled differently, this time by simply ignoring whether or not responses extend beyond the range $[0, L]$.

The starting point of this approach to bias sampling is to represent simple random sampling in terms of the Complementary Cumulative Distribution Function (CCDF), $\bar{F}(d) = P\{D > d\}$ of the duration D. Simulated values can be drawn by generating Uniform (0,1) realizations, U,

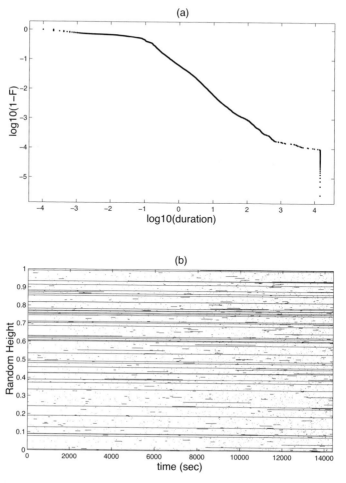

Fig. 10.6. *(a) Log-log transformed CCDF, showing how Uniform sampling will put more weight on larger responses. (b) Corresponding Mice and Elephants plot shows too much visual emphasis on the elephants.*

and then taking $D = \bar{F}^{-1}(U)$. Transformations of $\bar{F}$ will generate appropriate biased samples. The first example of this, based on a log-log transformation, is shown in Figure 10.6(a).

The horizontal axis in Figure 10.6 (a) shows the logarithm to base 10 of the duration, and the vertical axis is the logarithm to base 10 of the CCDF of the duration. This curve drives the sampling procedure as follows. First, select a random number $U \in [\log_{10}(1/N), 0]$ from the vertical axis and sample a corresponding response by choosing the maximum of the set

$$\{d_i : \log_{10} \bar{F}(d_i) \geq U\}, \tag{10.4}$$

where d_i, $i = 1, 2, \ldots, N$, represent the durations of the nonzero responses. Second, select another random number from the vertical axis
and choose a corresponding response by (10.4). If the selected response
is already in the sample, reject it and repeat the random sampling. Repeating these steps until there are 5,000 responses completes the sample.
Figure 10.6(b) shows a Mice and Elephants plot of such a sample.

Figure 10.6(b) shows too many elephants because the longer responses
have a much higher probability of being chosen, because the log transformation puts very large weight there. This is the opposite of random sampling which obscured the elephants, because the untransformed CCDF
put too little weight on them.

A family of compromises between these extremes is a sampling method
based on the Box–Cox family of transformations (Box and Cox; 1964) of
the CCDF of the durations. This family of transformations is defined as

$$G(d) = \frac{\bar{F}^\lambda(d) - 1}{\lambda \log 10}, \quad 0 \le \lambda \le 1. \tag{10.5}$$

The sampling procedure, after transforming the CCDF by (10.5), is the
same as above. Note that, if $\lambda = 0$, the sampling is the same as the one
with a log_{10} transformation, and if $\lambda = 1$, the sampling is consistent with
random sampling. The results of applying one member of this family of
transformations is shown in Figure 10.7.

Figure 10.7(a) shows the transformed CCDF of the duration with
$\lambda = 0.6$. The horizontal axis is still the log_{10} of the duration. This shows
that this Box–Cox transformation applies more weight to the intermediate size responses than to the elephants, compared to the log_{10} transformation.

Figure 10.7(b) shows the corresponding Mice and Elephants plot of the
selected 5,000 responses sampled using the transformed CCDF in Figure 10.7(a). Several very long elephants appear, but they are fewer than
in Figure 10.7(b) and the picture keeps many medium-sized responses.

The choice of $\lambda = 0.6$ was made on the basis of visual impression. Similar views, for a wide range of λ values, are available in a movie, which
is accessible in the Box–Cox Biased Sampling section of the webpage
González-Arévalo et al. (2004). The movie shows an increasing number
of elephants in the plots as λ decreases from 1 to 0. The general choice of
λ is an interesting open problem.

10.4 Quantile Window Sampling

Quantile window sampling provides a completely different view of the
HTTP responses. The main idea is to display the HTTP responses grouped
according to their lengths of duration. This visualization enables us to focus on the structure of various subgroups of the mice and elephants that

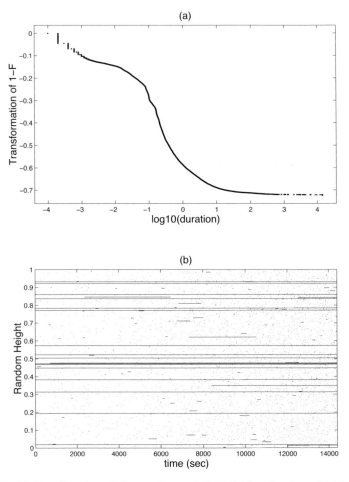

Fig. 10.7. *(a) Box–Cox* $\lambda = 0.6$ *transformed CCDF, showing how Uniform sampling will distribute weights among response durations. (b) Corresponding Mice and Elephants plot shows improved visual weighting.*

are similar in duration. The population of responses is divided into consecutive subsets of size 5,000, according to duration. That is, the longest 5,000 responses are displayed, the next longest 5,000 responses and so on. It is interesting to study the succession of these displays. A movie of them can be found in the Quantile Window Sampling section of the webpage González-Arévalo et al. (2004).

Figure 10.8 shows two such displays at the elephant end of the scale.

Figure 10.8(a) shows the biggest 5,000 elephants of the population. As expected, the visual impression is dominated by the long elephants crossing over most of the full four-hour time block. An interesting fact is that this group of responses contains a lot of mice as well. To investi-

(a)

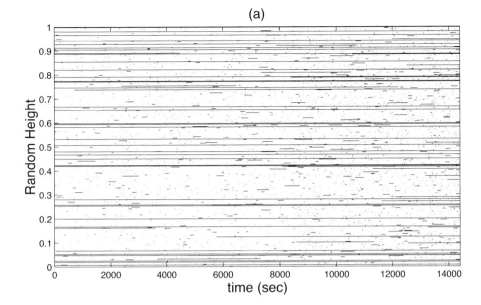

(b)

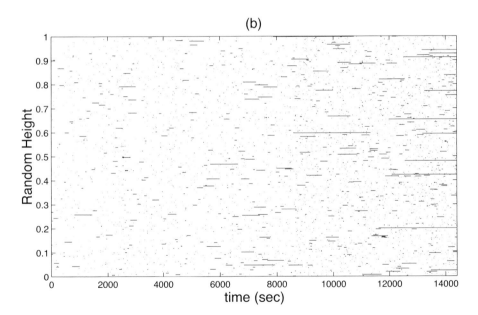

Fig. 10.8. *For the same Sunday morning HTTP responses: (a) The 5,000 biggest elephants. (b) The biggest 50 Elephants are removed and the next biggest 5,000 Elephants are displayed.*

gate this point further, the biggest 50 elephants were thrown out and the next biggest 5,000 elephants selected from the population. Figure 10.8(b) visualizes this sample. Only removing the biggest 50 elephants dramatically alters the visual impression, with now only medium-sized and short responses appearing. This provides a better view of other aspects of the population, including the fact that traffic level increases as the morning progresses, i.e., from left to right.

Figure 10.9 shows some lower quantile windows, chosen because they revealed some unexpected data artifacts.

Figure 10.9(a) shows 5,000 responses with durations from a minimum of 0.2146 to a maximum of 0.2262 (seconds). Basically, the responses shown in this figure have very similar duration lengths. An unexpected feature in this plot are some high-density vertical strips, representing bursts of responses of this duration at particular times. The biggest such burst appears around 9,500 (seconds) and that region is marked by two vertical lines. More bursts in terms of the starting times are shown in Figure 10.9 (b). This plot shows 5,000 responses with durations from a minimum of 0.6293 to a maximum of 0.7510 seconds. The biggest such vertical strip is observed around 8,000 (seconds) and that region is also marked by two vertical lines.

Insights into the causes of these bursts come from zooming in on the region between the red bars, and enhancing the visualization, as done in Figure 10.10. Additional useful information for understanding the causes of these bursts is the size of each HTTP response. Hence, in Figure 10.10, instead of using a random vertical height for separation, the size (in bytes) is used as the vertical axis.

The top panel of Figure 10.10(a) contains only the responses within the two vertical lines in Figure 10.9(a). Three big clusters are observed, with quite small response sizes and these are enclosed by a box. A further zoomed plot within this box is shown in the lower panel of Figure 10.10(a). Around 9,400, 9,410, and 9,430 (seconds), several small size responses are transferred practically at the same time, which creates the vertical strip that is seen in Figure 10.9(a). Deeper investigation revealed that these responses were composed of two packets and came from a common server, suggesting the download of a number of small webpage components, such as embedded thumbnail images. The three clusters have a timing that suggests they represent three such webpages. Usually such transfers involve single packets, and thus are not highlighted in the current analysis, because single packet responses have 0 duration. The reason that these responses come as two packets appears to be that the server sent the header of the response object in one packet, and then, after some extra processing, sent the actual data. HTTP attaches an application-level header to each response, and it seems that the particular server sending those responses is splitting header and data and also forcing TCP to send the header as soon as it is ready. This uncommon behaviour has some per-

(a)

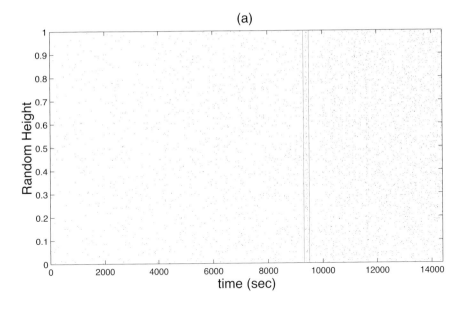

(b)

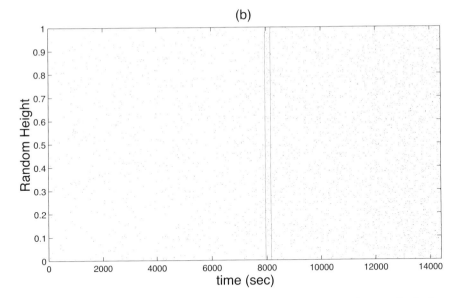

Fig. 10.9. *Two plots displaying different quantiles of the population. Both show some vertical bursts in terms of starting times.*

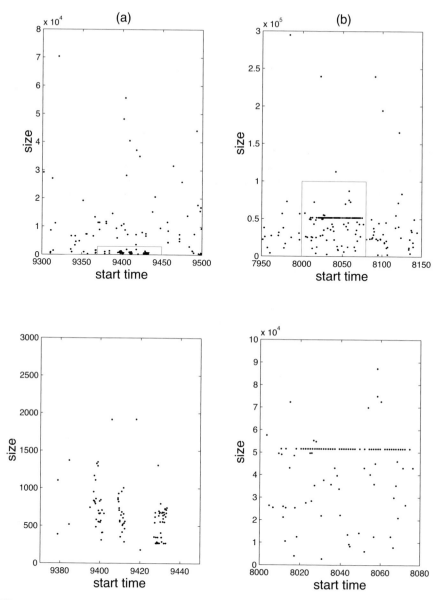

Fig. 10.10. *The burst phenomena within two vertical lines in Figure 10.9 are further investigated by zooming in on the highlighted regions. The vertical axis in these plots is the size of HTTP response, which aids the analysis. The left column shows zooms of Figure 10.9(a), and the right column shows zooms of Figure 10.9(b). The bottom row is a further zoom in on the rectangles drawn in the top row.*

formance benefits. The main benefit is that the client can start processing the header right away, rather than waiting for the header plus the object to arrive, slightly reducing the delay perceived by the user.

The top panel of Figure 10.10(b) shows another type of burst. This plot contains only the responses within the two vertical lines in Figure 10.9(b). This burst of HTTP responses is far different from the burst of embedded small objects discovered in Figure 10.10(a). This time, it is seen that the burst of responses all have essentially the same size. This strong similarity of size is confirmed by zooming in on the rectangular box, as shown in the lower panel. These responses are large enough to be sent as multiple packets. Deeper investigation showed that these all came from the same server, which was an internet game site. These appear to be a sequence of updates of game states, which explains the common size.

10.5 Commonality of Flow Rates

Another interesting view of the HTTP response data analysed in Section 10.2 is a scatterplot of the duration of each response against the size of the response in bytes (i.e., the amount of data transferred). Both variables share the heavy tailed *mice and elephants* behaviour demonstrated in Figures 10.1-10.3, so a reasonable view of the data comes from plotting both variables on the log scale. Figure 10.11 shows the resulting scatterplot. This requires special handling of responses with 0 duration (for instance for single packet responses) and they were dropped from the sample, which resulted in the 382,127 responses appearing in the scatterplot.

The general tendency in Figure 10.11 is roughly what one might expect: larger size responses need more time, so there is a general upward trend. Horizontal lines at the bottom of the plot reflect discreteness of very small time measurements. An initially surprising feature is the group of diagonal lines of points at larger times and sizes. Not only do the lines appear to be parallel, they also lie at a $45°$ angle to the coordinate axes, as indicated by the parallel dashed green line with equation $y = x - 2$. These diagonal lines of points represent sets of flows where

$$\log_{10} time = \log_{10} size + C,$$

for some constant C, which is the same as

$$size = R \cdot time,$$

where $R = 10^{-C}$ is interpretable as a constant rate. Thus, the flows following each diagonal line have essentially the same rate (defined as total size divided by total time).

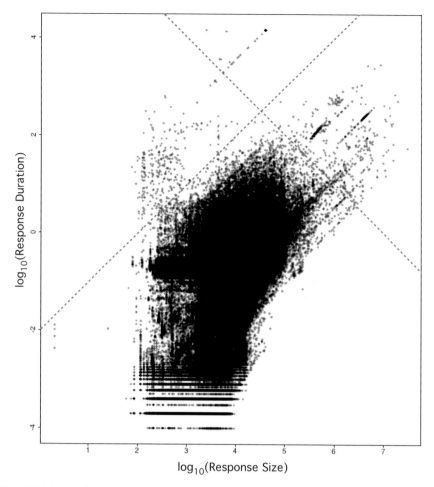

Fig. 10.11. *Log-log scatterplot showing how flow duration time depends on transmission size. Suggests clusters of flows with same throughput (rates) as diagonal lines.*

Figure 10.11 gives a strong visual impression that the large flows may be naturally clustered in terms of rates. This is sensible because rates are naturally driven by the nature of the network between the source and the destination. Most of the computers within UNC will likely have quite similar rates for a few popular websites, resulting in similar rates for large numbers of transfers.

The second main open problem of this chapter is to develop methods for analysing this aspect of the population. How can the clusters be isolated? What are the cluster boundaries? How many flows are in the major clusters?

A start on addressing these issues appears in Figure 10.12. The data are projected onto the orthogonal solid green line in Figure 10.11, so the problem is reduced to studying clusters in univariate data. For easy visual connection to Figure 10.11, the data are transformed to the coordinate system which treats the solid green line as the axis. The transformation is

$$proj = -2 - \frac{\log_{10} time - \log_{10} size}{\sqrt{2}}.$$

The denominator of $\sqrt{2}$ makes the transformation length invariant (i.e., a rotation), and the subtraction from -2 gives the most straightforward view of the solid green line as an axis. In particular, the solid green line is rotated in a counterclockwise fashion to give a conventional horizontal axis (as in Figure 10.12), and the points on the upper left part of the green line become negative values in the univariate view.

The top panel of Figure 10.12 shows two displays of the projected data. In the first, the green dots have horizontal coordinates of their projection values (i.e., the locations of the data points when projected onto the green line), and random vertical coordinates for visual separation. This jittered plot shows only a random sample of 10,000 to avoid overplotting problems. The second display of the data is the family of blue curves. These are kernel density estimates (essentially smooth histograms), with a wide range of window widths. Looking at a family of smooths is the scale space view of data, which is recommended as a practical solution to the traditional problem of bandwidth choice, see Chaudhuri and Marron (1999) for further discussion.

The family of kernel smooths suggests a number of broad bumps, and there is also a number of small spikes. It is tempting to dismiss the spikes as spurious sampling variability, but recall that clusters were suggested in Figure 10.11, and a possible physical explanation was suggested above. Furthermore the sample size $N = 382,127$ is fairly large, so perhaps those spikes represent important underlying structure in the data?

A useful tool for addressing such exploratory data analysis questions is the SiZer map shown in the bottom panel of Figure 10.12. Rows of this map correspond to different window widths, i.e., to blue kernel smooths, and the horizontal axis is the same as in the top panel. Colours are used to indicate statistical significance of the slopes of the blue curves, with blue (red) for significantly increasing (decreasing), with purple for regions where the slope is not significantly different from 0, and with gray where the data are too sparse for reliable inference.

The SiZer map shows that all of the broad bumps are statistically significant, as are most of the tall thin bumps. These may not be surprising because $N = 382,127$ allows resolution of quite a few features of the underlying probability density, in view of the large sample size. More surprising may be the very small bump at -1.7. This is hardly visible in the blue family, and yet is clearly statistically significant in the SiZer map.

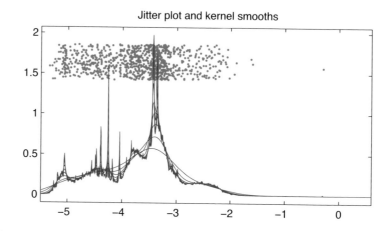

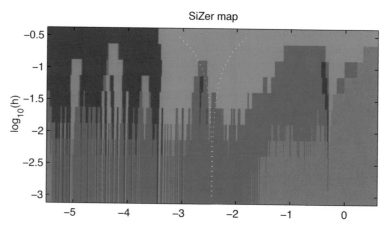

Fig. 10.12. *The SiZer analysis of the projected scatterplot shows many significant clusters.*

The analysis of Figure 10.12 is not very satisfying, because it seems that perhaps some of the clusters, which are clearly visible as lines of points in Figure 10.11, might be masked by the large amount of other data that makes up the broad peaks. A simple approach to this is to repeat the analysis for a suitably thresholded sub-sample. Visual inspection of Figure 10.11 suggests using only the data above the solid green line, $y = -x + 7$. There were 572 such points, still enough for effective kernel density estimation. The resulting analysis is shown in Figure 10.13.

As expected, the broad bumps in Figure 10.12 have now disappeared. There are also some very significant slim spikes. Note that the spike near -1.7 is now much taller in the blue family of curves (essentially all of these data points have been retained from Figure 10.12 and are now pro-

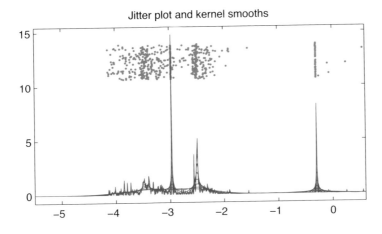

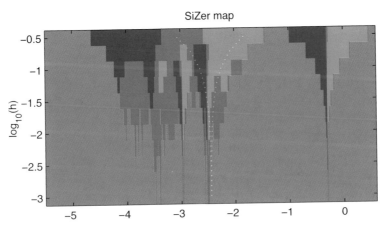

Fig. 10.13. *SiZer analysis of scatterplot points above the solid green line in Figure 10.11. It shows significant clusters, but different from those of Figure 10.12. Many more clusters than those in Figure 10.11 exist.*

portionally a far larger part of the population). However, note that many of the tall thin peaks in Figure 10.12 are not present in Figure 10.13. This shows that much of the clustered aspect of the population actually occurs more in the main body of the main scatterplot in Figure 10.11 and thus cannot be teased out by simple thresholding as done in Figure 10.13.

This is a case where the scatterplot of Figure 10.11 hides a large amount of interesting population structure. The SiZer analysis is an indirect way of understanding this. A more direct way of visualizing this type of structure is to use α-blending as described in Section 3.5. Whereas the scatterplot in Figure 10.11 uses an α-value of 0.5, the α-value in Figure 10.14 is as low as 0.025, i.e., just 1/20 of the original value. Much of

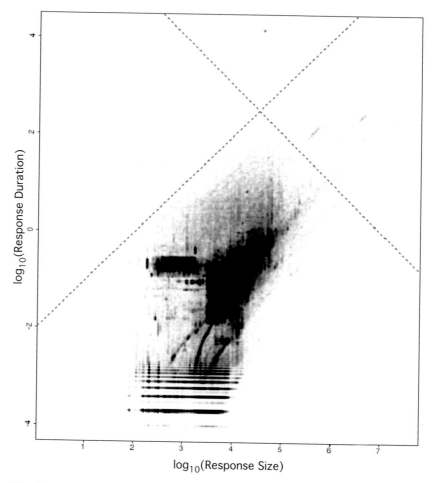

Fig. 10.14. *The same data as in Figure 10.11, now with an α-value of 0.025.*

the structure revealed in Figures 10.12 and 10.13 can also be seen here. The figure also suggests that some hidden structure might be uncovered with further exploration, whether with zooming in and other α-values or with additional SiZer analyses.

11

Graphics of a Large Dataset

Antony Unwin and Martin Theus

Wir reden zuviel, wir sollten weniger sprechen und mehr zeichnen.[1]

Johann Wolfgang von Goethe, to Johannes Daniel Falk, 14 June 1809

11.1 Introduction

Describing visualization methods is one thing, putting them into action another. In the final chapter, a real dataset is analysed to illustrate the contribution visualization can make in practice. The dataset of High-Tech company information was used for the InfoVis 2005 conference contest (Grinstein et al.; 2005, see `http://ivpr.cs.uml.edu/infovis05`). Obviously the application here concentrates on what visualization can do, though many analytical tools could contribute as well. Information may be uncovered in the data using a variety of approaches, and the graphics described are just some of many that could be drawn. Other graphics may highlight the same features in different ways or could emphasise other aspects of the dataset. There are many, many possibilities. Every data analysis is different and so there are no hard and fast rules.

The next section gives a brief summary of the major stages of a data analysis and the chapter roughly follows this ordering. Data analysis is not a linear process with a fixed sequence of predefined steps from beginning to end, but some form of structure should be kept in mind. At all times, it is valuable to be able to talk to people who know how the data were collected, what the variables mean, and which conclusions are important.

For an introduction consider Figure 11.1, a parallel coordinates plot of the aggregate sales of companies by State over the 15 years of the dataset. It gives a good summary of the overall trend. The general rise is partly due to inflation and partly due to the growth of High-Tech industries. Linking to a map (see Section 11.3.6) can help to interpret the apparent clustering.

---

[1] We talk too much, we should speak less and draw more.

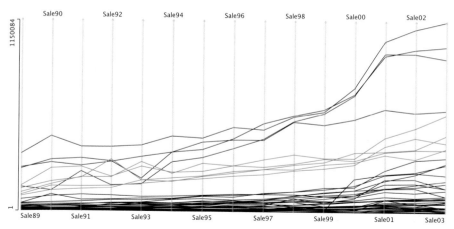

Fig. 11.1. *Parallel coordinate plot of the aggregate sales of companies by State over the fifteen years of the dataset. In 1998, there is a clear separation into three groups: a top group consisting of CA, NY, TX, and MI; a second group (selected) consisting of CT, NJ, PA, OH, and IL; and the rest. Over the course of the next 5 years, MI could not keep up with the growth of CA, NY, and TX and fell back into the second group.*

11.2 QuickStart Guide
Data Visualization for Large Datasets

1. **Preliminaries**
   Find out the background
   – what are the goals of the study?
   – what are the data? (numbers of variables and cases)
   Prepare the data
   – formatting, storing
   – merging and matching
2. **Variables**
   What types of data?
   – nominal, ordinal, continuous, spatial, ...
   What properties do the data have?
   – individual plots
   – range, categories, missings, outliers
   Assess the data quality and clean the data
   Check for consistency and plausibility
3. **First analyses**
   Use multiple basic plots together
   – barcharts, histograms, scatterplots
   Explore the plots using interaction
   – especially querying, linking, reformatting
   Summarise low-dimensional properties of the data

4. Multivariate displays

Choose subsets of variables relevant to the goals

Draw multivariate plots

- parallel coordinates, scatterplot matrices for continuous
- mosaic plot variations for categorical variables

Explore using interaction, including sorting and zooming

Identify any higher dimensional structure

5. Grouping and selection

Condition on grouping variables and compare subsets

- boxplots by one grouping variable
- small multiples, trellis displays

Consider interesting selected subsets in isolation (drill-down)

6. Special features

Reduce dataset

- aggregate data by groups or clusters (case reduction)
- combine related variables (dimension reduction, e.g., PCA)

Use any temporal structure (time series plots)

Link to maps if there is spatial information

7. Presenting results

Choose and polish graphics that convey the main conclusions

Ensure that all graphics are fully explained

11.3 Visualizing the InfoVis 2005 Contest Dataset

11.3.1 Preliminaries

Background

The aims and background information about the data were provided on a webpage (which is an uncommon situation for an analysis, as usually the connection between the analyst and the dataset provider is much closer). Questions of interest included characterizing clusters of industries and identifying unusual companies and regions.

Data Preparation

The data were made available in a collection of datasets, two for each year plus one with variable information and one with the geographic locations of zipcodes. There were data for 85,000 firms over 15 years. One set of files gave information by firm and the other gave information by firm activities (for instance, in 2003 there were some 57,000 firms and just under 200,000 activities). It was possible to

- combine the data in one or more larger datasets
  (which makes them readily accessible to statistics software);

- set up a relational database
  (which makes it easy to reorganize them in different forms).

11.3.2 Variables

Types of Data

In the raw InfoVis datasets, there were nine variables per year over the 15 years: an ID variable, three for the address (city, state, and zipcode), one categorical (Industry sector), one structured categorical (NAICS, an industrial classification code), one discrete (year of founding), and two continuous (sales and numbers of employees).

11.3.3 First Analyses

Categorical Variables

Categorical variables may be nominal or ordinal. Industry sector and State clearly have no numerical ordering and were treated as nominal. NAICS codes are numerical but should not be treated as ordinal. Their structure is hierarchical, so the digits do contain information, they are not just nominal values. The same goes for zipcodes.

There are 18 Industry sectors and 52 States (including Puerto Rico and Washington, DC). The upper panel of Figure 11.5 shows how the companies were distributed by Industry and State in 2003. There were large numbers of both Software and Telecommunications companies. Amongst the States, California dominated, with Massachusetts, New York, and Texas some way behind.

With large datasets, simple categorical variables may have very many categories — an issue that is discussed in Section 3.3.2. In this application, there was a variable called primary NAICS code, which in the first

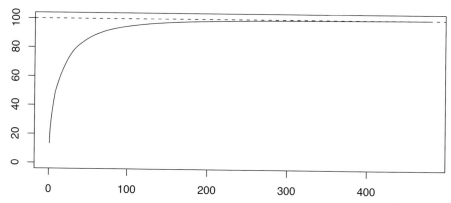

Fig. 11.2. *Cumulative shares of sales in 2003 by NAICS categories.*

year of the study in 1989 had 142 categories and in the final year, 2003, 479. Needless to say, many categories were very small. The biggest 59 categories in 1979 contained 90% of the categories and in 2003 it was the biggest 65 categories (out of 479!) that contained 90%. Categories with very low frequencies could be ignored, combined, or perhaps highlighted as interesting, unusual cases. For the InfoVis dataset it is worth remembering that aggregate sales may be a better measure of importance than numbers of companies. The largest 70 categories in 2003 amount to 90% of sales (Figure 11.2), but only 79% of the companies. One solution was to aggregate the NAICS codes across their first two digits. This left 24 groups with 6 still being relatively small.

Continuous Variables

Figure 11.3 is a missing value plot for each year of sales data. The numbers of companies rose steadily till 2001, but declined in the following two years.

Patterns of missing values for sets of variables can be looked at in two ways. In Figure 11.4, the upper plot displays missing values for 5 variables from 2002 individually. Because the data for many companies are missing altogether in each year, the combinations of missings for companies with at least some data are more interesting and these are displayed in the fluctuation diagram in the lower plot of Figure 11.4, which was generated from the missing value plot. The largest group amongst the companies with both some values present and some missing has been selected, about 5% of all the companies.

It is not so much the number of companies, but the sales they achieve. The barcharts in the lower panel of Figure 11.5 are weighted by sales in 2003 and show how the importance of the

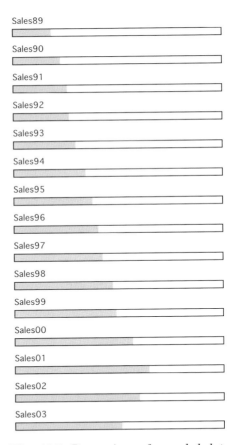

Fig. 11.3. *Proportions of recorded data and missing values for all companies for the years 1989 to 2003.*

NON Industry sector ("not primarily High-Tech") increases substantially

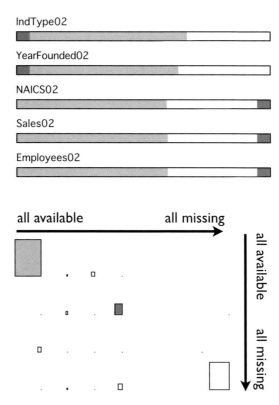

Fig. 11.4. *A missing value plot of the five major variables for the year 2002 (above). The 4301 cases missing on Sales, Employees, and NAICS, but with values for Ind-Typ and YearFounded have been selected. A fluctuation diagram for the five 2002 variables showing which combinations of missings arise (below). Most cases are all present (grey cell top left) or all absent (white cell bottom right).*

when sales and not just counts are considered, while that of the Telecommunications group declines. Energy is also much more important in terms of sales. In the barchart of States, the relative positions of California and New York change, and Massachusetts is now no longer even in the top ten. Comparing the barcharts in the upper and lower panels of Figure 11.5 can be done for major differences, but not for more precise comparisons. Alternatives include side-by-side versions, as in Figure 11.6. On the left, the original ordering is retained and on the right, the sectors have been sorted by numbers of companies. The Energy sector stands out even more and the low values for Manufacturing and Factory Automation become apparent.

Both sales and employee figures were highly skew. Figure 11.7 shows the employee numbers distribution for 2003, both for the raw data (left) and for log transformed data (right). Redmarking is obviously needed in

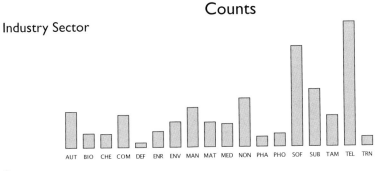

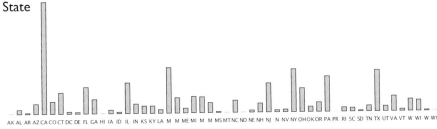

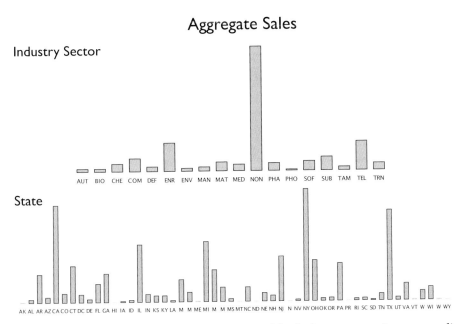

Fig. 11.5. *Numbers of companies by State and by Industry sector (upper panel) Aggregate sales by State and by Industry sector in 2003 (lower panel).*

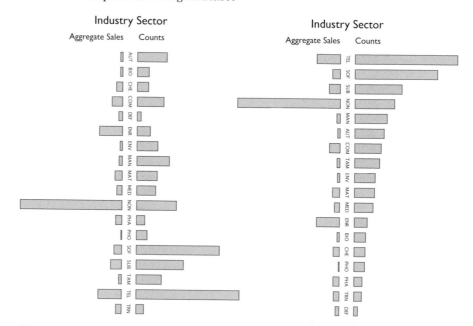

Fig. 11.6. *Companies by Industry sector, both by numbers and by aggregate sales, ordered alphabetically on the left and by numbers of companies on the right.*

the raws plot and reveals that one outlier group on its own (in fact, just one company) extends the scale by a factor of two. Redmarking is also useful in the logged histogram. You could look for modes, gaps, clusters, or accumulations at favoured values. Summary statistics (maximum, minimum, mean, median, and possibly others) will reveal other aspects of the variables. For instance, in the InfoVis dataset, there were a few sales figures that were negative. Large datasets are like that.

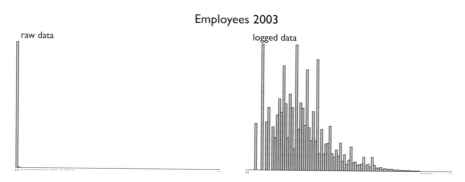

Fig. 11.7. *Distribution of employee numbers in 2003 for 45,000 companies, raw data (left) and log transformed (right).*

11.3.4 Multivariate Displays

Pairs of Continuous Variables

There is a suggestion in the right-hand plot of Figure 11.7 that some values occur particularly often. A scatterplot of log Sales and log Employee numbers, Figure 11.8 (top), offers supporting evidence, and varying the level of α-blending, the effect becomes obvious, Figure 11.8 (bottom). It does not come as a big surprise that company figures are often reported in rounded numbers.

Some outliers can be very informative. After time series of the sales data showed a few strange excessive peak patterns (Figure 11.10), scatterplots like the one in Figure 11.9 were drawn to check how serious the problem might be. Sales jumping from 0 to a high value and back again could just about be plausible (a very successful start-up disappearing through takeover), but high sales falling to a much lower non-zero figure after a sharp rise look suspicious. The plot shows two extreme outliers top left, companies with few sales in 1992 and very many in 1993, and two bottom right, companies with strong sales in 1992 but hardly any in 1993.

In this scatterplot, there is so much overplotting along the diagonal that there is no method that can accurately represent the data density. On the other hand, this makes the cases not on the diagonal all the more interesting. Plots like this led to investigating year on year changes and to the discovery that for a high proportion of companies the same sales figures were recorded for both years (39.5% of those operating in both 1989 and 1990 with sales greater than 1 in both years and 47.1% of the 29,500 companies operating in both 2002 and 2003 with sales greater than 1 in both years). Serendipitous discoveries are a feature of exploratory work.

Pairs of Categorical Variables

Two categorical variables can be plotted in mosaic plots. Fluctuation diagrams are best for displaying year on year transitions (e.g., Industry sectors in two successive years as in Figures 11.12 and 11.13). Same binsize plots are useful for drawing attention to combinations that never occur and for comparing proportions across groups irrespective of their sizes, e.g., Figure 11.16 for State by Year.

Combinations of More Than Two Continuous Variables

Missings can be a problem in multivariate displays. In the InfoVis dataset, this difficulty was avoided for sales and employee figures by replacing missings with zeros. You could say that if a firm does not exist, then it has neither sales nor employees.

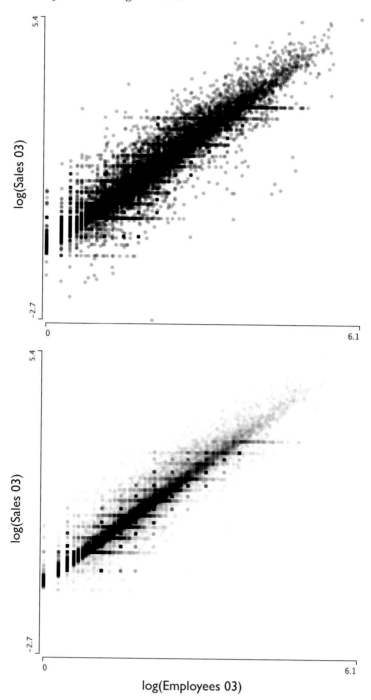

Fig. 11.8. *Log Sales in 2003 plotted against log Employee numbers in 2003 with a default value for α-blending (top), and with a higher level of α-blending (bottom).*

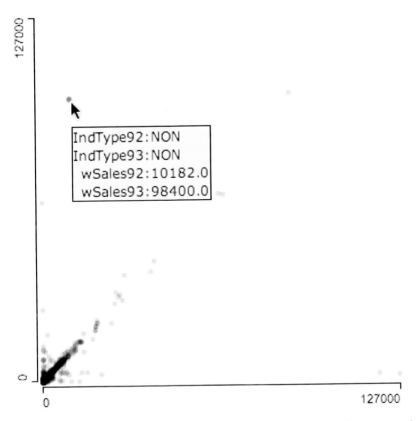

Fig. 11.9. *Sales in 93 plotted against sales in 92. The scales are the same on both axes. One of the outliers has been queried.*

Parallel coordinate plots are a good way of gaining an overview of many continuous variables. If it makes sense to scale all of a group of variables to a common scale, then the plot is even more informative, e.g., the yearly sales figures of the companies in Figure 11.10. Alpha-blending (see Section 3.5) may be used to counteract overplotting with large numbers of cases, as with the scatterplots in Figure 11.8.

(Dimension reduction methods, like principal components, *may* be useful with large numbers of continuous variables. However, the effects of outliers and skew distributions should be dealt with first, otherwise standardising of individual variables carried out prior to dimension reduction will produce poor results.)

Combinations of More Than Two Categorical Variables

Fluctuation diagrams of Industry by State show where industries are located (look down the columns) and which industries are important for

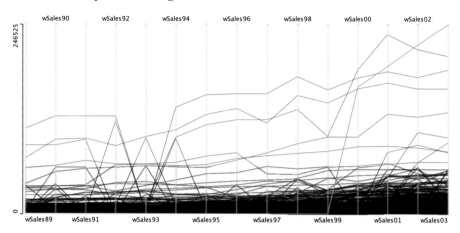

Fig. 11.10. *Parallel coordinate plot for all 72,148 companies with non-zero sales.*

the States (look across the rows). The diagram on the left of Figure 11.11 displays numbers of companies in 2003. The main feature is the large numbers of companies in several sectors in California. The diagram on the right of Figure 11.11 is weighted by aggregate sales in 2003. The column for the Industry sector NON stands out, as does the cell towards the lower left of the plot representing the sector Energy in Texas. California is less important by sales than it is by numbers of companies, something that was already obvious in Figure 11.5.

The sharp rise of sales in the Industry sector NON could have been due to reclassification of large companies. The upper plot of Figure 11.12 shows a fluctuation plot of Industry sector in 2000 (rows) and 2001 (columns). Ceiling-censored zooming has been used to downplay the strong diagonal effect due to most companies not changing sector. The rightmost column shows the numbers of companies in the 2000 dataset that were not in the 2001 dataset, and the bottom row shows the companies that appeared in 2001 but not in 2000. The large numbers of new companies in the sectors NON and Telecommunications is striking. In particular, there were more new companies in the NON sector in 2001 than in that sector altogether in 2000.

The Industry sector change plot can be weighted with sales in 2000, emphasizing what companies sold before they changed sector (off-diagonal elements) or disappeared (final column). This is shown in the lower plot of Figure 11.12 where the companies moving to NON from Environment make up the largest group changing sector. In Figure 11.13, the weighting is by sales in 2001, emphasizing what companies sold after they moved sectors (off diagonal cells) or what companies new to the dataset sold (bottom row). Here it is the high sales by the "new" companies in the NON sector that stand out.

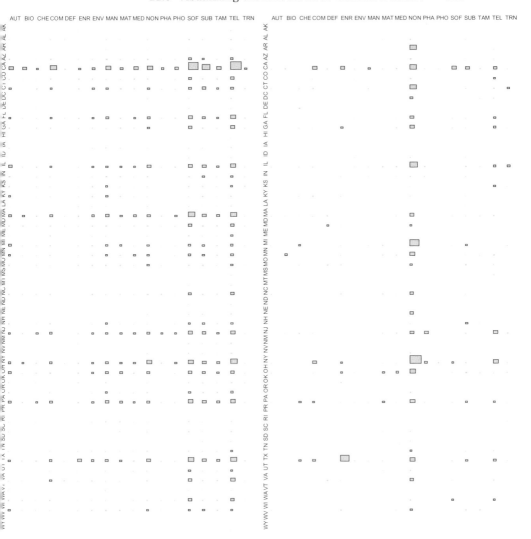

Fig. 11.11. *A fluctuation diagram of companies by State and Industry, raw counts (left), weighted by Sales in 2003 (right).*

11.3.5 Grouping and Selection

Selecting Industry sectors in a barchart and linking to sales/employee numbers scatterplots like Figure 11.8 reveals different data patterns. The 900 chemical companies follow a common linear relationship, whereas the 1,100 energy companies are more varied with more companies having relatively higher sales (see Figure 11.14).

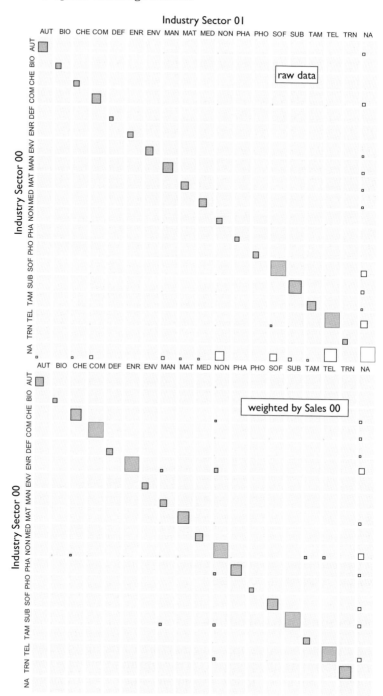

Fig. 11.12. *Ceiling-censored fluctuation plot of Industry sectors in 2000 and 2001 (top): same plot weighted by sales in 2000 (bottom).*

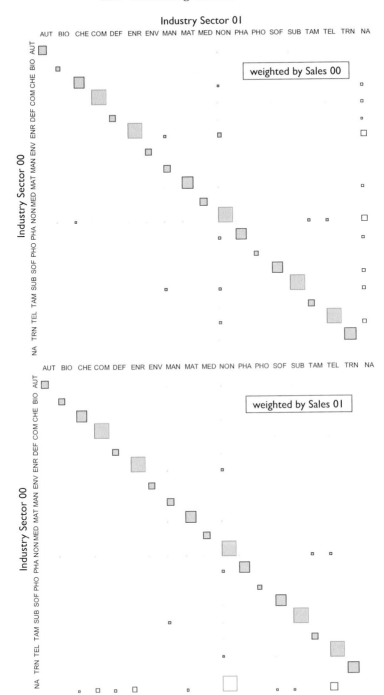

Fig. 11.13. *Ceiling-censored fluctuation plot of Industry sectors in 2000 and 2001 weighted by sales in 2000 (top): same plot weighted by sales in 2001 (bottom).*

Sorting barcharts in different ways can also shed light on individual groups. Figure 11.15 shows three displays for the States in 2003. The top plot is ordered by numbers of companies in California and demonstrates that California is like the US as a whole, at least in terms of relative numbers of companies. The middle plot is ordered by sales of Californian companies and shows which sectors are important to California. The NON sector is most important, but others are too, and notice especially the Energy sector. The third plot is ordered by the proportion of sales by Californian companies and shows in which sectors California is important for the nation. Here the sectors Software and Communication stand out, while NON and Energy are ranked far lower.

Figure 11.16 is a same binsize mosaic plot, which can be used to judge how strong Industry sectors are in different States over the years. The Chemical Industry has been selected, and its dominance in Delaware is clearly visible, as is its declining importance in Wyoming. In this view, all bins are the same size, so highlighting is comparing proportional values, not absolute values. The bins are weighted by sales. Obviously, the absolute figures would look quite different.

The State and Industry sectors together give $52 \times 18 = 936$ combinations. Other factors of interest include the year the firm was founded. Concentrating on newer firms, those founded since 1988, would give $52 \times 18 \times 15 = 14,040$ combinations, far too many to distinguish patterns. Concentrating on the biggest three States by total sales reduces the number of combinations to a manageable 1350 as in Figure 11.17. California is represented by the cells to the left, New York by those in the middle, and Texas by those on the right. Noteworthy are the groups of firms from individual years with high sales, for instance the 5 Energy companies founded in Texas in 2002 (the largest cell on the right), and the 22 Telecommunications companies founded in 2001 in New York (the largest cell in the bottom row). By contrast, the fairly regular pattern of Californian firms in Telecommunications (the cells to the left in the bottom row) is the exception rather than the rule. (The details for these graphics were obtained by querying the individual cells.)

11.3.6 Special Features

The temporal structure in the InfoVis dataset is simple, one value per variable per year for 15 years. It leads to large numbers of short time series, but none of the difficulties you can face with comparing series of different lengths, recorded at different, irregular time points. Plotting the 80,000-plus series together, as in Figure 11.10, was done using parallel coordinates because it offered the option of displaying boxplots by timepoints to compare distributions by year (which is useful for identifying outliers).

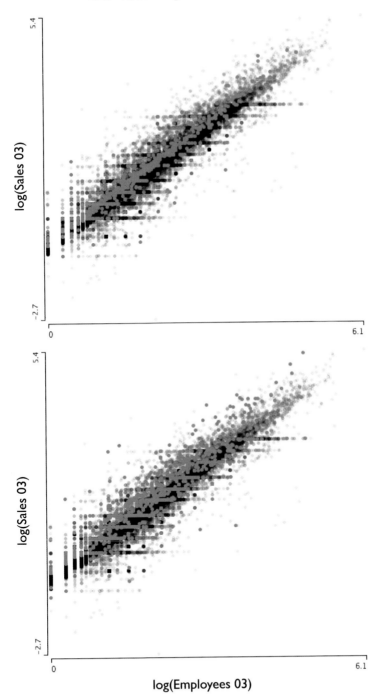

Fig. 11.14. *Log Sales in 2003 plotted against log Employee numbers in 2003 with the Chemical sector highlighted (top) and the Energy sector highlighted (below).*

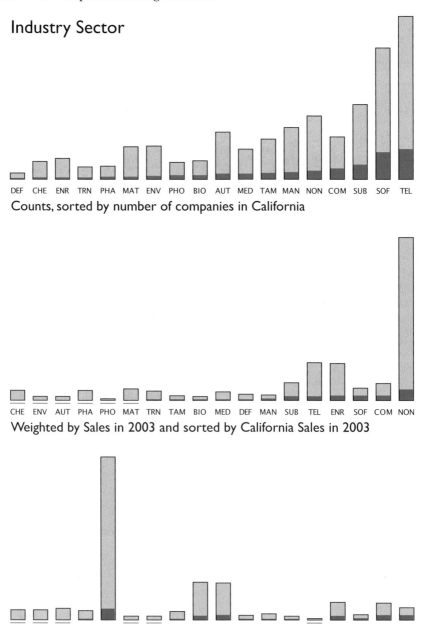

Fig. 11.15. *Three different ways of sorting Industry sectors using sales information from companies in California.*

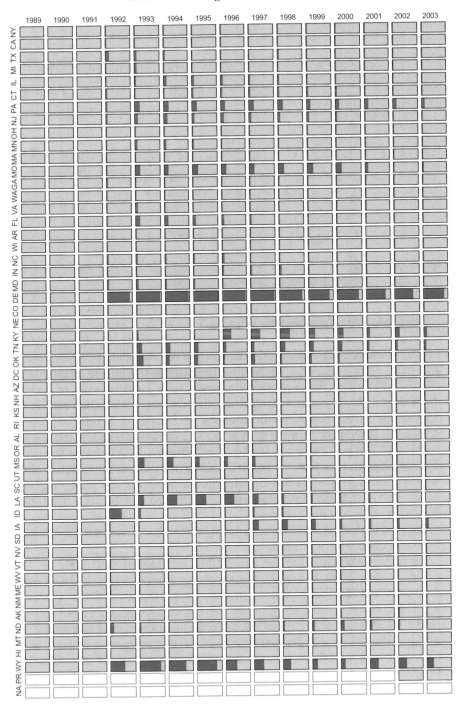

Fig. 11.16. *A Mosaic Plot of State by Year weighted by sales in the 'Same Binsize' view with the Chemicals Industry (CHE) group highlighted. States sorted by size.*

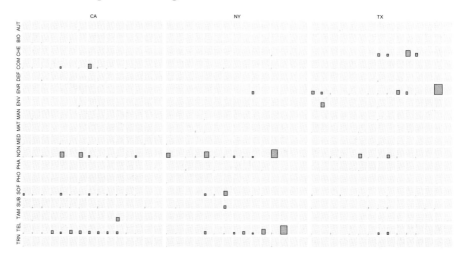

Fig. 11.17. *Fluctuation diagram by Industry and Year Founded (after 1988) for California, New York, and Texas, weighted by Sales in 2003.*

The dataset included spatial information by State and by zipcode. The State information could be used to produce displays like Figure 11.19 showing the US States coloured according to aggregated 2003 sales data. California, New York, and Texas clearly stand out. The file of longitude and latitude values for each zipcode produced a location map of points, Figure 11.18, highly concentrated on the Coasts with some relatively empty areas in the centre of the country. To construct a more representative view, the zipcodes were converted to counties using the databases available at http://www.zipinfo.com.[2] Figure 11.20 shows the 2003 sales data again, this time by county, and reveals much more detail than was possible in Figure 11.19. Consistent patterns for the New England area as well as for the Bay Area and Los Angeles can be seen. Counties with zero values are not coloured.

Maps on both State and County level can be downloaded from US governmental sites free of charge.

Scatterplots of zipcodes for the 10,613 companies that were in the dataset in both 1989 and 2003 are shown in Figure 11.21. A little over half (5482) had the same zipcode in both years, those on the main diagonal. The left scatterplot shows that there was a broad spread of moves. Depending on your opinions of High-Tech companies, you can either be surprised that such a high percentage of firms has moved or that such a high percentage of firms has not moved. Further ways of investigating this phenomenon include fluctuation diagrams for moves between States,

---
[2] Several public domain zipcode-databases turned out to be too inaccurate to be useful.

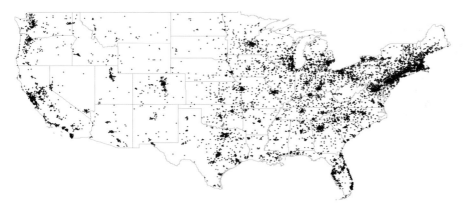

Fig. 11.18. *Zipcode locations of companies, based on their last recorded zipcode (excluding companies in Alaska, Hawaii, and Puerto Rico).*

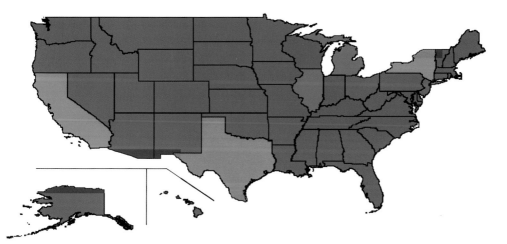

Fig. 11.19. *A choropleth map of the US States coloured by sales in 2003, ranging from light blue (low) to bright red (high).*

plots using the distance between zipcodes, and maps linked to these displays. There are a lot of alternatives — as there always are with graphics.

11.3.7 Presenting Results

The graphics in this chapter have been drawn with several software packages to illustrate the range of possibilities available. Although they present some interesting findings, they have been drawn primarily for exploratory purposes not so much for presentation purposes. Analysing large datasets involves drawing lots of graphics, not just one or two.

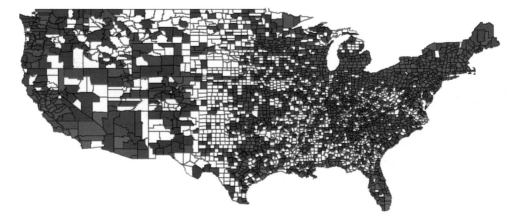

Fig. 11.20. *A choroplethmap of the US Counties coloured by sales in 2003. The same colour scale has been used as in Figure 11.19, and counties with zero values are not coloured.*

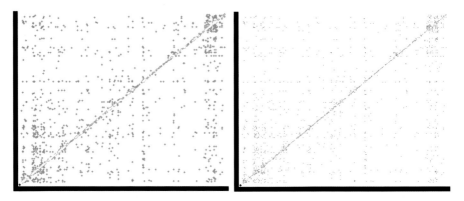

Fig. 11.21. *Zipcode locations of companies in the dataset for both 1989 and 2003. The plot on the right uses a smaller point size to emphasise the diagonal of firms with unchanged zipcodes.*

Even if an analysis is quite narrowly focused on a specific question, large datasets offer plenty of views and aspects to explore.

More attention could have been paid to framing, scaling, colour, aspect ratio, annotations, and all the other components that make up the aesthetics of graphics. There are some rules that should always be respected, but many options are a subjective matter to a great extent. Readers can mull over what choices they themselves would have made for individual displays. "Perfect" solutions are rare.

Up till now, presentation of graphics has meant static paper printed plots. This is going to change, and interactive and flexible displays will become available via the web. One notable example is the US Cancer Mortality Atlas, which can be found at `http://www.cancer.gov/atlasplus/`. With all that flexibility to hand, it is important to remember that the correct interpretation is mostly left to the reader of the graphic displays.

11.3.8 Summary

Visualization is excellent for displaying overall trends (like the increasing sales in Figure 11.1, the patterns of sales by State in Figures 11.5 and 11.6, and the distribution of employee numbers in Figure 11.7), for revealing individual details (like the odd sales jumps noticed in Figure 11.10 and confirmed in Figure 11.9), for uncovering data quality issues (like the missing patterns in Figure 11.4, the rounding of data observed in Figure 11.8, and the sales that were the same in successive years, initially suggested by Figure 11.9), for comparing groups (like the different industry patterns in Figure 11.14), for checking local patterns in various ways (like that of the Industry sectors in California in Figure 11.15), for identifying influential subsets (like the new companies in the NON group in Figure 11.12), for comparing the importance of different groups (like the patterns of Industry by State such as in Figures 11.11, 11.12, and 11.16), for looking at selected subsets (like the development of new companies in the biggest States in Figure 11.17), for revealing spatial patterns (like in Figures 11.18 to 11.20), and for looking at changes over time (like the time series for companies in Figure 11.10 and the scatterplots of zipcode changes in Figure 11.21).

Every large dataset is unique and special, and there are many different and alternative ways of looking at the data. To include all the graphics that depict informative features of the InfoVis dataset would require many more pages than have been used here. The aim has been to give the flavour of how visualization can be used to uncover information in data and to illustrate the variety of questions and the range of information that come to light in exploring a big dataset.

Now it is time for readers to draw graphics for large datasets for themselves . . .

References

Adams, R. and Wu, M. (2002). Pisa 2000 technical report, *Technical report*, OECD.

Andrews, D. (1972). Plots of high dimensional data, *Biometrics* **28**(2): 125–136.

Asimov, D. (1985). The grand tour a tool for viewing multidimensional data, *SIAM Journal on Scientific and Statistical Computing* **6**(1): 128–143.

Becker, R. A. and Cleveland, W. S. (1987). Brushing Scatterplots, *in* Cleveland and McGill (1988), pp. 201–224.

Becker, R. A., Cleveland, W. S. and Shyu, M.-J. (1996). The Visual Design and Control of Trellis Displays, *Journal of Computational and Graphical Statistics* **6**(1): 123–155.

Bertin, J. (1983). *Semiology of Graphics*, 2nd edn, University of Wisconsin Press, Madison.

Bolton, R., Hand, D. J. and Crowder, M. (2004). Significance tests for unsupervised pattern discovery in large continuous multivariate data sets, *CSDA* **46**(1): 57–79.

Box, G. P. E. and Cox, D. R. (1964). An analysis of transformations, *Journal of the Royal Statistical Society, Series B* **26**: 211–243.

Braverman, A. (2002). Compressing massive geophysical datasets using vector quantization, *JCGS* **11**(1): 44–62.

Breiman, L. (1996). Bagging predictors, *Machine Learning* **24**(2): 123/140.

Breiman, L. (1999). Random forests - random features, *Technical Report TR567*, University of California Berkeley, Statistics Department.

Breiman, L., Friedman, J. H., Olshen, R. and Stone, C. (1984). *Classification and Regression Trees*, Wadsworth.

Buja, A., Asimov, D., Hurley, C. and McDonald, J. A. (1988). Elements of a viewing pipeline for data analysis, *in* W. S. Cleveland and M. McGill (eds), *Dynamic Graphics for Statistics*, Wadsworth, Belmont, CA, pp. 277–308.

Buja, A., Cook, D., Asimov, D. and Hurley, C. (1996). Theory and computational methods for dynamic projections in high-dimensional data visualization.
URL: *http://www.research.att.com/areas/stat/xgobi/papers/dynamic-projections.ps.gz*

Buja, A. and Tukey, P. (eds) (1991). *Computing and Graphics in Statistics*, Springer-Verlag, New York, NY.

Cao, J., Cleveland, W. S., Lin, D. and Sun, D. X. (2001). On the nonstationarity of internet traffic, *Proceedings of the ACM SIGMETRICS '01*, pp. 102–112.

Cao, J., Cleveland, W. S., Lin, D. and Sun, D. X. (2002a). The effect of statistical multiplexing on the long range dependence of internet packet traffic, *Technical report*, Bell Labs Tech Report.

Cao, J., Cleveland, W. S., Lin, D. and Sun, D. X. (2002b). Internet traffic: Statistical multiplexing gains, *DIMACS Workshop on Internet and WWW Measurement, Mapping and Modeling*.

Cao, J., Cleveland, W. S., Lin, D. and Sun, D. X. (2002c). Internet traffic tends toward poisson and independent as the load increases, *in* C. Holmes, D. Denison, M. Hansen, B. Yu, B. M. C. Holmes, D. Denison, M. Hansen, B. Yu and B. Mallick (eds), *Nonlinear Estimation and Classification*, Springer, New York.

Cappelli, C., Mola, F. and Siciliano, R. (2002). A statistical approach to growing a reliable honest tree, *Computational Statistics and Data Analysis* (38): 285–299.

Card, S. K., MacKinlay, J. D. and Shneiderman, B. (1999). *Readings in Information Visualization: Using Vision to Think*, Morgan Kaufmann Publishers, Inc., San Franciscon, CA.

Carr, D. B. (1985). Evaluation of Graphical Techniques for Data in Dimensions 3 to 5: Scatterplot Matrix, Glyph, and Stereo Examples, *ASA Proceedings of the Section on Statistical Computing*, American Statistical Association, Alexandria, VA, pp. 229–235.

Carr, D. B. (1991). *Looking at Large Data Sets Using Binned Data Plots*, in Buja and Tukey (1991), pp. 7–39.

Carr, D. B., Littlefield, R. J., Nicholson, W. L. and Littlefield, J. S. (1987). Scatterplot matrix techniques for large n, *JASA* **82**(398): 424–436.

Carr, D. B., Olsen, A. R., Courbois, J.-Y. P., Pierson, S. M. and Carr, D. A. (1998). Linked micromap plots: Named and described, *Statistical Computing & Statistical Graphics Newsletter* **9**(1): 24–32.

Carr, D. B., Wegman, E. J. and Luo, Q. (1997). ExplorN: Design considerations past and present, *Technical Report 137*, Center for Computational Statistics, George Mason University, Fairfax, VA.

Carr, D. B., Zhang, Y. and Li, Y. (2002). Dynamically conditioned choropleth maps, *Statistical Computing & Statistical Graphics Newsletter* **13**(2): 2–7.

Chambers, J. (1967). Some general aspects of statistical computing, *Applied Statistics* **16**(2): 124–132.

Chambers, J. (1999). Computing with data: Concepts and challenges, *The American Statistician* **53**(1): 73–84.

Chaudhuri, P. and Marron, J. S. (1999). Sizer for exploration of structure in curves, *Journal of the American Statistical Association* **94**: 807–823.

Chipman, H., George, E. and McCulloch, R. (1998). Bayesian cart model search (with discussion), *Journal of the American Statistical Association* **93**(443): 935–960.

Cho, K. (2001). *Aguri Traffic Density Graph*.
   URL: *http://www.csl.sony.co.jp/person/kjc/kjc/software/aguri-density.html*

Clemons, T. and Pagano, M. (1999). Are babies normal?, *The American Statistician* **53**(4): 298–302.

Cleveland, W. S. (1994). *The Elements of Graphing Data*, revised edn, Hobart Press, Summit, NJ.

Cleveland, W. S. and McGill, M. E. (1984). Graphical perception: Theory, experimentation and application to the development of graphical methods, *Journal of the American Statistical Association* **79**(387): 531–554.

Cleveland, W. S. and McGill, M. E. (eds) (1988). *Dynamic Graphics for Statistics*, Wadsworth, Monterey, CA.

Coale, A. J. and Stephan, F. F. (1962). The case of the indians and the teen-age widows, *JASA* **57**(298): 338–347.

Coleman, M. K. (1996). Aesthetics-based graph layout for human consumption, *Software Practice and Experience* **26**(12): 1415–1438.

Cook, D., Buja, A., Cabrera, J. and Hurley, C. (1995). Grand Tour and Projection Pursuit, *Journal of Computational and Graphical Statistics* **4**(3): 155–172.

Cox, D. R. (1984). Long-range dependence: A review, *in* H. A. David and H. T. David (eds), *Statistics: An Appraisal, Proceedings of the 50th Anniversary Conference*, The Iowa State University Press, pp. 55–74.

Crovella, M. E. and Bestavros, A. (1996). Self-similarity in world wide web traffic evidence and possible causes, *Proceedings of the ACM SIGMETRICS 96*, Philadelphia, PA, pp. 160–169.

Davidson, R. and Harel, D. (1996). Drawing graphs nicely using simulated annealing, *ACM Transactions on Graphics* **15**(4): 301–331.

Dawson, R. J. M. (1995). The "unusual episode" data revisited, *Journal of Statistics Education* **3**(3).

Di Battista, G., Eades, P., Tamassia, R. and Tollis, I. (1994). Algorithms for drawing graphs: An annotated bibliography, *Computational Geometry* (4): 235–282.

Dimsdale, B. (1984). Conic transformations and projectivities, *Technical report*, IBM Los Angeles Scientific Center, Santa Monica, CA.

Downey, A. B. (2000). The structural cause of file size distributions, *Technical report*, Wellesley College Tech. Report CSD-TR25-2000.
URL: *http://rocky.wellesley.edu/downey/filesize/*

Downey, A. B. (2001). Evidence for long tailed distributions in the internet, *ACM SIGCOMM Internet Measurement Workshop*.
URL: *http://rocky.wellesley.edu/downey/longtail/*

Duffield, N. G. and Lund, C. (2003). Predicting resource usage and estimation accuracy in an ip flow measurement collection infrastructure, *ACM SIGCOMM Internet Measurement Conference 2003*.

DuMouchel, W., Volinsky, C., Johnson, T., Cortes, C. and Pregibon, D. (1999). Squashing files flatter, *KDD-99 Proceedings*, pp. 6–15.

Dykes, J., MacEachren, A. and Kraak, M.-J. (2005). *Exploring Geovisualization*, Elsevier.

Eddy, W. F., Fitzgerald, M., Genovese, C., Lazar, N., Mockus, A. and Welling, J. (1999). The challenge of functional magnetic resonance imaging, *JCGS* **8**(3): 545–558.

Eick, S. G. and Karr, A. F. (2002). Visual scalability, *Journal of Computational and Graphical Statistics* **11**(1): 22–43.

Eick, S. G. and Wills, G. (1993). Navigating large networks with hierarchies, *Proceedings of IEEE Visualization '93*.

Fayyad, U., Grinstein, G. and Wierse, A. (eds) (1999). *Information Visualization in Data Mining and Knowledge Discovery*, Morgan Kaufmann.

Fisher, R. A. (1991). *Statistical methods, experimental design, and scientific inference*, Oxford University Press, Oxford.

Fisherkeller, M. A., Friedman, J. H. and Tukey, J. W. (1971). *PRIM-9: An Interactive Multidimensional Data Display and Analysis System*, in Cleveland and McGill (1988).

Friedman, J. H. (2001). The role of statistics in the data revolution, *International Statistical Review* **69**(1): 5–10.

Friendly, M. (1994a). Mosaic displays for multi-way contingency tables, *Journal of the American Statistical Association* **89**: 190–200.

Furnas, G. W. (1986). Generalized fisheye views, *SIGCHI '86 Conference on Human Factors in Computing Systems*, ACM, pp. 16–23.

Galton, F. (1892). *Hereditary Genius*, second edn, MacMillan.

Gehrke, J. E., Ganti, V. and Ramakrishnan, R. (2000). RainForest - a framework for fast decision tree construction of large datasets, *Data Mining and Knowledge Discovery* **4**: 127–162.

Gelman, A. (2004). Exploratory data analysis for complex models, *JCGS* **13**(4): 755–779.

Gong, W., Liu, Y., Misra, V. and Towsley, D. (2001). On the tails of web file size distributions, *Proceedings of 39-th Allerton Conference on Communication, Control, and Computing*.
URL: *http://www-net.cs.umass.edu/networks/publications.html*

González-Arévalo, B., Hernández-Campos, F., Marron, J. S. and Park, C. (2004). *Visualization Challenges in Internet Traffic Research*.

URL: *http://www.cs.unc.edu/Research/dirt/proj/marron/Visual Challenge/*

Good, I. J. and Gaskins, R. A. (1980). Density estimation and bump-hunting by the penalized likelihood method exemplified by scattering and meteorite data, *JASA* **75**(369): 42–56.

Grinstein, G., Cvek, U., Derthick, M. and Trutschl, M. (2005). *IEEE Info-Vis 2005 Contest, Technology Data in the US*.
   URL: *http://ivpr.cs.uml.edu/infovis05*

Grosjean, J., Plaisant, C. and Bederson, B. (2002). Spacetree: Supporting exploration in large node link tree, design evolution and empirical evaluation, *Procedings of IEEE Symposium on Information Visualization*, pp. 57–64.

Han, Y. (2000). *Visualization of weighted data*, PhD thesis, Department of Statistics, Iowa State University.

Hand, D. J., Blunt, G., Kelly, M. G. and Adams, N. M. (2000). Data mining for fun and profit, *Statistical Science* **15**(2): 111–131.

Hannig, J., Marron, J. S. and Riedi, R. (2001). Zooming statistics: Inference across scales, *Journal of the Korean Statistical Society* **30**(2): 327–345.

Harris, B. (1959). Further remarks on a data plotting machine, *The American Statistician* **14**(4): 38+46.

Hartigan, J. A. and Kleiner, B. (1981). Mosaics for contingency tables, *13th Symposium on the Interface*, Springer Verlag, New York, pp. 268–273.

Haslett, J., Bradley, R., Craig, P., Unwin, A. R. and Wills, G. (1991). Dynamic Graphics for Exploring Spatial Data with Application to Locating Global and Local Anomalies, *The American Statistician* **45**(3): 234–242.

Haughton, D., Deichmann, J., Eshghi, A., Sayek, S., Teebagy, N. and Topi, H. (2003). A review of software packages for data mining, *American Statistician* **57**(4): 290–309.

Healy, M. J. R. (1952). Some statistical aspects of anthropometry, *JRSS B* **14**(2): 164–184.

Heath, D., Resnick, S. and Samorodnitsky, G. (1998). Heavy tails and long range dependence in on/off processes and associated fluid models, *Mathematics of Operations Research* **23**: 145–165.

Herman, I. and Marshal, M. (2000). Graphxml - an xml-based graph description format, *Proceedings of the 8th International Symposium on Graph Drawing*, pp. 52 – 62.

Hernández-Campos, F., Marron, J. S., Samorodnitsky, G. and Smith, F. D. (2002). Variable heavy tailed durations in internet traffic, *Performance Evaluation* .
   URL: *http://www.cs.unc.edu/Research/dirt/proj/marron/VarHeavy Tails/*

Hochheiser, H. and Shneiderman, B. (2004). Dynamic query tools for time series data sets, timebox widgets for interactive exploration, *Information Visualization* **3**(1): 1–18.

Hofmann, H. (2000). Exploring categorical data: interactive mosaic plots, *Metrika* **51**(1): 11–26.

Hofmann, H. and Wilhelm, A. (2001). Visual comparison of association rules, *Computational Statistics* **16**(3): 399–415.

Hohn, N. and Veitch, D. (2003). Inverting sampled traffic, *ACM/SIG-COMM Internet Measurement Conference*, pp. 222–233.

Hothorn, T. and Lausen, B. (2003). Double-Bagging: Combining classifiers by bootstrap aggregation, *Pattern Recognition* **36**(6): 1303–1309.

Huang, C., McDonald, J. A. and Stuetzle, W. (1997). Variable Resolution Bivariate Plots, *Journal of Computational and Graphics Statistics* **6**(4): 383–396.

Huber, P. J. (1992). Issues in computational data analysis, *in* Y. Dodge and J. Whittaker (eds), *Computational Statistics*, Vol. 2.

Huber, P. J. (1994). Huge data sets, *Compstat '94: Proceedings, eds. Dutter, R., Grossmann, W., Physica, Heidelberg.*

Huber, P. J. (1999). Massive datasets workshop: Four years after, *JCGS* **8**(3): 635–652.

Hummel, J. (1996). Linked bar charts: Analysing categorical data graphically, *Computational Statistics* **11**: 23–33.

Inselberg, A. (1985). The Plane with Parallel Coordinates, *The Visual Computer* **1**: 69–91.

Kahn, R. and Braverman, A. (1999). What shall we do with the data we are expecting from upcoming earth observation satellites?, *JCGS* **8**(3): 575–588.

Karagiannis, M. F. T., Molle, M. and Broido, A. (2004). A nonstationary poisson view of internet traffic, *Proceedings of IEEE Infocom*.

Kehoe, C. M., Pitkow, J. and Morton, K. (1997). Gvu's eighth www user survey report (in color), *Research report*, Georgia Tech Research Corporation, Atlanta, GA.

Keim, D. (2000). Designing pixel-oriented visualization techniques: Theory and applications, *IEEE Transactions on Visualization and Computer Graphics* **6**(1): 59–78.

Kettaneh, N., Berglund, A. and Wold, S. (2005). Pca and pls with very large data sets, *CSDA* **48**(1): 69–85.

Kettenring, J. (2001). Massive data sets ... reflections on a workshop, *Computing Science and Statistics, Proceedings of the 33rd Symposium on the Interface*, Vol. 33, Interface Foundation, Montreal.

Kraak, M.-J. and Ormeling, F. J. (1996). *Cartography — Visualization of spatial data*, Longman.

Lamping, J. and Rao, R. (1995). The hyperbolic browser: A focus+context technique for visualizing large hierarchies, *in* S. K. Card, J. D. MacKinlay and B. Shneiderman (eds), *Readings in Information Visualization: Using Vision to Think*, Morgan Kaufmann, chapter 4, pp. 382–408.

Leland, W. E., Taqqu, M. S., Willinger, W. and Wilson, D. V. (1994). On the self-similar nature of ethernet traffic (extended version), *IEEE/ACM Transactions on Networking* **2**: 1–15.

Mandelbrot, B. B. (1969). Long-run linearity, locally gaussian processes, h-spectra and infinite variance, *International Economic Review* **10**: 82–113.

Marchette, D. J. and Wegman, E. J. (2004). Statistical analysis of network data for cybersecurity, *Chance* **17**(1): 8–18.

Marron, J. S., Hernández-Campos, F. and Smith, F. D. (2002). Mice and elephants visualization of internet traffic, *in* W. Härdle and B. Rönz (eds), *Proceedings of the CompStat 2002*, Springer, Berlin, pp. 47–54. **URL:** *http://www.cs.unc.edu/Research/dirt/proj/marron/Mice Elephants/*

McDonald, J. A. (1988). Interactive graphics for data analysis, *in* W. S. Cleveland and M. McGill (eds), *Dynamic Graphics for Statistics*, Wadsworth.

McIntosh, A. (1999). Analyzing telephone network data, *JCGS* **8**(3): 611–619.

McNeill, D. R. and Tukey, J. W. (1975). Higher-order diagnosis of two-way tables, *Biometrics* **31**(2): 487–510.

Murrell, P. (2005). *R Graphics*, Chapman & Hall, London.

Nielsen, J. (1994). *Usability Engineering*, Morgan Kaufmann, San Francisco.

Nievergelt, J. and Hirichs, K. (1993). *Algorithms and Data Structures with Applications to Graphics and Geometry*, Prentice Hall, Englewood Cliffs, NJ.

Pach, J. (ed.) (2004). *Graph Drawing, 12th International Symposium*, Lecture Notes in Computer Science, Springer, New York.

Page, E. S., Healy, M. J. R., Jeffers, J. N. R., Baines, A. and Paddle, G. M. (1967). Discussion (of statistical programming), *Applied Statistics* **1967**(2): 133–148.

Playfair, W. (2005). *Playfair's Commercial and Political Atlas and Statistical Breviary*, Cambridge, London.

Porter, R. (1891). The eleventh census, *JASA* **2**(15): 321–379.

Posse, C. (2001). Hierarchical model-based clustering for large datasets, *Journal of Computational and Graphical Statistics* **10**(3): 464–486.

Post, F., Nielson, G. and Bonneau, G.-P. (eds) (2002). *Data Visualization: The State of the Art*, Kluwer.

Putnam, R. D. (2000). *Bowling Alone*, Touchstone, New York.

Rao, R. and Card, S. K. (1994). The Table Lens: Merging graphical and symbolic representations in an interactive focus context visualization for tabular information, *Proc. ACM Conference on Human Factors in*

Computing Systems, CHI, ACM.
    URL: *citeseer.ist.psu.edu/rao94table.html*
Reed, W. J. and Jorgensen, M. (2004). The double pareto-lognormal distribution - a new parametric model for size distribution, *Communications in Statistics -Theory and Methods* **33**: 1733–1753.
Resnick, S. and Samorodnitsky, G. (1999). Activity periods of an infinite server queue and performance of certain heavy tailed fluid queues, *Queueing Systems* **33**: 43–71.
Ripley, B. D. (2005). How computing has changed statistics, *in* A. Davison, Y. Dodge and N. Wermuth (eds), *Celebrating Statistics: Papers in Honour of Sir David Cox on His 80th Birthday*, Oxford U.P., pp. 197–212.
Robertson, G. G., MacKinlay, J. D. and Card, S. K. (1991). Cone trees: animated 3d visualizations of hierarchical information, *CHI '91: Proceedings of the SIGCHI conference on Human factors in computing systems*, ACM Press, pp. 189–194.
Ruggles, S. and Sobek, M. (1997). *Integrated Public Use Microdata Series: Version 2.0*, Minneapolis: Historical Census Projects, University of Minnesota.
    URL: *http://www.ipums.org*
Rundensteiner, E. A., Ward, M. O., Yang, J. and Doshi, P. R. (2002). Xmdvtool: visual interactive data exploration and trend discovery of high-dimensional data sets, *SIGMOD '02: Proceedings of the 2002 ACM SIGMOD international conference on Management of data*, ACM Press, New York, NY, pp. 631–631.
Schapire, R. E. (1999). A brief introduction to boosting, *16th International Joint Conference on Artificial Intelligence*, Morgan Kaufmann Publishers Inc., San Francisco, CA.
Scott, D. (1992). *Multivariate Density Estimation — Theory, Practice, and Visualization*, Wiley, New York, NY.
Shneiderman, B. (1996). The eyes have it: A task by data type taxonomy for information visualizations, *VL '96: Proceedings of the 1996 IEEE Symposium on Visual Languages*, IEEE Computer Society, p. 336.
Smith, F. D., Hernández-Campos, F., Jeffay, K. and Ott, D. (2001). "What TCP/IP protocol headers can tell us about the web", *Proceedings of ACM SIGMETRICS 2001/Performance 2001*, pp. 245–256.
Spence, R. (2000). *Information Visualization*, Addison Wesley, Harlow, England.
Stolte, C., Tang, D. and Hanrahan, P. (2002). Multiscale visualization using data cubes, *Proceedings of InfoViz '02, IEEE Symposium on Information Visualization*, IEEE Computer Society Press.
Student (1931). The lanarkshire milk experiment, *Biometrika* **23**(3/4): 398–406.

Swayne, D. F., Cook, D. and Buja, A. (1991). Xgobi: Interactive graphics in the x window system with a link to S, *American Statistical Association: Proceedings of the Section on Statistical Graphics*.

Swayne, D. F. and Klinke, S. (1999). Editorial, *Computational Statistics* **14**(1).

Swayne, D. F., Lang, D. T., Buja, A. and Cook, D. (2003). GGobi: evolving from XGobi into an extensible framework for interactive data visualization, *Computational Statistics and Data Analysis* **43**(4): 423–444.

Taqqu, M. S. and Levy, J. (1986). *Progress in probability and statistics*, Birkhaeuser, Boston, chapter Using renewal processes to generate LRD and high variability, pp. 73–89.

Theus, M. (1996). *Theorie und Anwendung interaktiver statistischer Graphik*, Augsburger Mathematisch-Naturwissenschaftliche Schriften, Band 14, Wissner, Augsburg.

Theus, M. (2002a). Geographical information systems, *Handbook of Data Mining and Knowledge Discovery*, Oxford University Press.

Theus, M. (2002b). Interactive Data Visualization using Mondrian, *Journal of Statistical Software* **7**(11).

Theus, M., Hofmann, H. and Wilhelm, A. (1998). Selection sequences — interactive analysis of massive data sets, *in* D. Scott (ed.), *Proceedings of the 29th Symposium on the Interface: Computing Science and Statistics*, pp. 439–444.

Theus, M. and Lauer, S. (1999). Visualizing loglinear models, *Journal of Computational and Graphical Statistics* **8**(3): 396–412.

Thomas, J. and Cook, K. (eds) (2005). *Illuminating the Path: The Research and Development Agenda for Visual Analytics*, IEEE Press.

Tierney, L. (1991). *LispStat: An Object-Oriented Environment for Statistical Computing and Dynamic Graphics*, Wiley, New York, NY.

Tufte, E. R. (1983). *The Visual Display of Quantitative Information*, Graphic Press, Cheshire, Connecticut.

Tukey, J. W. and Tukey, P. (1985). Computer graphics and exploratory data analysis: An introduction, *Sixth Annual Conference and Exposition: Computer Graphics*, National Computer Graphics Association, pp. 773–785.

Tukey, J. W. and Tukey, P. (1990). Strips displaying empirical distributions: Textured dot strips, *Technical report*, Bellcore Technical Memorandum.

Turo, D. and Johnson, B. (1992). Improving the visualization of hierarchies with treemaps: Design issues and experimentation, *3rd IEEE Conference on visualization*.

Unwin, A. R. (1999). Requirements for interactive graphics software for exploratory data analysis, *Journal of Computational Statistics* **1**: 7–22.

Unwin, A. R., Hawkins, G., Hofmann, H. and Siegl, B. (1996). Interactive graphics for data sets with missing values - manet, *JCGS* **5**(2): 113–122.

Unwin, A. R., Volinsky, C. and Winkler, S. (2003). Parallel coordinates for exploratory modelling analysis, *Computational Statistics & Data Analysis* **43**(4): 553–564.

Unwin, A. R., Wills, G. and Haslett, J. (1990). REGARD - Graphical Analysis of Regional Data, *ASA Proceedings of the Section on Statistical Graphics*, American Statistical Association, Alexandria, VA, pp. 36–41.

Urbanek, S. and Theus, M. (2003). iPlots — High Interaction Graphics for R, *Proceedings of the DSC 2003 Conference*.

Velleman, P. F. (1997). *DataDesk Version 6.0 — Statistics Guide*, Data Description Inc., Ithaca, NY.

Venables, W. N. and Ripley, B. D. (1999). *Modern Applied Statistics with S+*, 3rd edn, Springer, New York.

Ward, M. O., Peng, W. and Wang, X. (2004). Hierarchical visual data mining for large-scale data, *Computational Statistics* **19**(1): 147–158.

Wegman, E. J. (1990). Hyperdimensional Data Analysis Using Parallel Coordinates, *Journal of American Statistics Association* **85**: 664–675.

Wegman, E. J. (1991). The Grand Tour in k-Dimensions, *Technical Report 68*, Center for Computational Statistics, George Mason University.

Wegman, E. J. (1995). Huge data sets and the frontiers of computational feasibility, *Journal of Computational and Graphical Statistics* **4**(4): 281–295.

Wegman, E. J. (2003). Visual data mining, *Statistics in Medicine* **22**: 1383–1397 + 10 color plates.

Wegman, E. J. and Carr, D. B. (1993). Statistical graphics and visualization, *in* C. R. Rao (ed.), *Handbook of Statistics*, Vol. 9, North Holland, Amsterdam, pp. 857–958.

Wegman, E. J., Poston, W. L. and Solka, J. L. (1998). Image Grand Tour, *Automatic Target Recognition VIII - Proceedings of SPIE, 3371*, SPIE, Bellingham, WA, pp. 286–294. Republished, Vol 6: Automatic Target Recognition. The CD-ROM, (Firooz Sadjadi, ed.), SPIE: Bellingham, WA, 1999.

Wegman, E. J. and Solka, J. L. (2002). On some mathematics for visualizing high dimensional data, *Sankhya* **64**(2): 429–452.

Wegman, E. J. and Solka, J. L. (2005). Statistical data mining, *Handbook of Statistics*, North Holland, Amsterdam, pp. 1–44.

Whittaker, J. (1990). *Graphical Models on Applied Multivariate Statistics*, John Wiley and Sons, New York.

Wilkinson, L. (2005). *The Grammar of Graphics*, 2nd edn, Springer, New York.

Wills, G. (1997). Visual exploration of large structured data sets, *New Techniques and Technologies for Statistics II*, IOS Press, Washington DC.

Wills, G., Unwin, A. R., Haslett, J. and Craig, P. (1990). Dynamic interactive graphics for spatially referenced data, *Fortschritte der Statistik-Software 2*, Gustav Fischer Verlag, Stuttgart, pp. 278–287.

Yates, F. (1966). Computers, the second revolution in statistics, *Biometrics* **22**(2): 233–251.

Zhang, Z.-L., Ribeiro, V. J., Moon, S. B. and Diot, C. (2003). Smalltime scaling behaviors of internet backbone traffic: an empirical study, *Proceedings of IEEE Infocom*.

Authors

264   Authors

Index

 **Springer**
the language of science

springeronline.com

The Grammar of Graphics
Second Edition

Leland Wilkinson

This book was written for statisticians, computer scientists, geographers, researchers, and others interested in visualizing data. It presents a unique foundation for producing almost every quantitative graphic found in scientific journals, newspapers, statistical packages, and data visualization systems. The second edition is almost twice the size of the original, with six new chapters and substantial revision. Much of the added material makes this book suitable for survey courses in visualization and statistical graphics.

2005. 694 p. (Statistics and Computing) Hardcover ISBN 0-387-24544-8

Pattern Recognition and Machine Learning

Christopher M. Bishop

The dramatic growth in practical applications for machine learning over the last ten years has been accompanied by many important developments in the underlying algorithms and techniques. This completely new textbook reflects these recent developments while providing a comprehensive introduction to the fields of pattern recognition and machine learning. It is aimed at advanced undergraduates or first-year PhD students, as well as researchers and practitioners. No previous knowledge of pattern recognition or machine learning concepts is assumed.

2006. 702 p. (Information Science and Statistics) Hardcover ISBN 0-387-31073-8

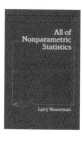

All of Nonparametric Statistics

Larry Wasserman

The goal of this text is to provide the reader with a single book where they can find a brief account of many, modern topics in nonparametric inference. This text covers a wide range of topics including: the bootstrap, the nonparametric delta method, nonparametric regression, density estimation, orthogonal function methods, minimax estimation, nonparametric confidence sets, and wavelets. The book has a mixture of methods and theory.

2005. 276 p. (Springer Texts in Statistics) Hardcover ISBN 0-387-25145-6

Easy Ways to Order▶ Call: Toll-Free 1-800-SPRINGER • E-mail: orders-ny@springer.com • Write: Springer, Dept. S8113, PO Box 2485, Secaucus, NJ 07096-2485 • Visit: Your local scientific bookstore or urge your librarian to order.